CATTLE, STRAW

AND

SYSTEM CONTROL

a study of straw feeding systems

Promotoren: **Dr. D. Zwart**
Hoogleraar in de Tropische Veehouderij

Dr. J.A. Renkema
Hoogleraar in de Agrarische Bedrijfseconomie

J.B. Schiere

CATTLE, STRAW

AND

SYSTEM CONTROL

a study of straw feeding systems

Proefschrift

ter verkrijging van de graad van
doctor in de landbouw- en milieuwetenschappen,
op gezag van de rector magnificus,
Dr. C.M. Karssen,
in het openbaar te verdedigen
op vrijdag 26 mei 1995
des namiddags te vier uur in de aula
van de Landbouwuniversiteit te Wageningen

Omslag: Wim Valen

CIP-DATA KONINKLIJKE BIBLIOTHEEK, DEN HAAG

Schiere, J.B.

Cattle, straw and system control : a study of straw feeding systems / J.B. Schiere.
- Amsterdam : Koninklijk Instituut voor de Tropen. - Ill.
Also publ. as thesis Landbouwuniversiteit Wageningen, 1995.
- With ref. - With summary in Dutch.
ISBN 90-6832-098-X
NUGI 835
Subject headings: straw feeding systems

Schiere, J.B., Cattle, Straw and System Control, a study of straw feeding systems (Vee, Stro en Systeembeheersing, een studie over strovoederingssystemen)

Straw is an important animal feed in many farming systems of the world. It can be fed in different ways, and for a variety of objectives. An analysis of the role of straw is therefore undertaken to explain the usefulness of straw feeding methods in different systems. Automatically this leads to the question about the role of straw in the drive and shape of farming systems. A review of backgrounds and approaches of Farming Systems Research (FSR) is given to serve as a background for the discussions. After that, this thesis proposes a classification of farming systems that reflects a form of system evolution with different degrees of system control. The classification also provides a framework that tentatively explains the usefulness of straw feeding methods over a large range of systems. Based on feeding trials and a review of literature, a set of feeding values and animal responses are obtained that help to understand the economics of some specific straw feeding methods: urea treatment of straw and/or supplementation with better feeds. It is shown that the feeding of urea treated straw is best feasible in conditions with a) medium levels of individual animal output in terms of milk and meat, b) limited access to better quality feeds and c) access to required inputs such as urea. The calculations are tested from a nutritional angle as well as from an approach that distinguishes between high and low input agriculture, here called closed and open systems. The results are used for thought experiments that explore the possibilities to adjust feed and animals in different combinations for maximum system output. This shows that mutual adjustment of resources and (sub)system production objectives is necessary for maximum system output. This conclusion agrees with concepts from thermodynamics, information theory and ecology, branches of science that also help to explain the drive and shape of systems.

Ph.D. thesis, Department of Animal Production Systems, and Department of Agricultural Economics, P.O.Box 338, 6700AH Wageningen, The Netherlands.

VOORWOORD

Na terugkeer in Nederland ben ik ongeveer in 1987 begonnen met denken over dit proefschrift, o.a. bedoeld om mijn tropenkennis meer algemeen toepasbaar te maken. Het begon met het feit dat ik vanuit Sri Lanka de beschikking had over gegevens van dierproeven betreffende het gebruik van stro als veevoer. Uiteindelijk is slechts een deel van de Sri Lankaanse gegevens gebruikt voor dit proefschrift (vooral in de secties 3 en 4). Hoe langer hoe meer is het een zoektocht geworden naar beter begrip van de wisselwerking tussen het gedrag van landbouwsystemen en de introduktie van technologieën. Een en ander was geïnspireerd door het werk aan het BIOCON project in India, deelname aan adviseringsmissies, onderwijs en het schrijven/editen van boeken en beleidsnota's. Soms gebeurde het schrijfwerk in een kantoor in Engeland of op een bovenkamer in Utrecht, afgesloten van de Wageningse omgeving. Het meeste denk- en leeswerk gebeurde in wachtkamers, in 'het veld' van verschillende landen, op reis in trein, vliegtuig of boot, in hotels en guesthouses, en thuis aan de keukentafel.

Bewust heb ik het verhaal de kans gegeven zich zelf te vormen, maar terugkijkend kan ik zeggen dat het geheel de karakteristieken heeft gekregen van een 'damning objective', een begrip dat in Hfdst. 5.2 wordt uitgewerkt. Uiteraard waren mijn eigen "resources" te beperkt voor deze klus. Vooral mijn vrouw en kinderen hebben het een en ander van nabij meegemaakt en mij gesteund. Daarnaast zijn er zoveel anderen, in binnen en buitenland, op hoge en lage posities die allen op hun eigen - vaak onvermoede - wijze bijdroegen. Jan de Wit heeft wellicht de meest direkte rol gespeeld met zijn fundamentele kritiek, filosofische kommentaar én hulp bij de konkrete uitwerking. Verder waren er natuurlijk vele andere mensen die kritische vragen stelden, suggesties deden of literatuur natrokken en aangaven. Hoewel 'native english speakers' hebben geholpen om taalfouten te verbeteren, ligt de schuld van lange zinnen en taalfouten bij mijzelf. De meeste reviewers van individuele hoofdstukken staan genoemd aan het eind van de betreffende hoofdstukken, maar speciale dank ben ik verschuldigd aan Gerard Oomen, Arend Jan Nell, en prof. Leen 't Mannetje voor de inhoudelijke gesprekken, en aan Erna Minten en Arno Maas voor hun geduldige advies en hulp bij het afwerken. Last but not least wil ik mijn promotoren bedanken, prof. Dick Zwart en prof. Jan Renkema, die zich wel eens zullen hebben afgevraagd wat voor orde er uit deze 'chaos' tevoorschijn zou komen. De collega's van de afdeling Dierlijke Produktie Systemen, en met name mijn nieuwe baas, prof. Herman van Keulen, worden bedankt voor hun begrip voor het feit dat twee maanden werk, toch nog wel twee jaar in beslag kunnen nemen.

Bennekom, Februari 1995

aan: Heit, Mem,
        Rinske, Irene, Conny, Marcus,
        Sarah N'Dipitee en Wong Cilik.

"you are right" replied Michelangelo with a sigh, "but I have not yet conceived the final dome. I shall have to find it"
        p. 745 in the paperback version of Irving Stone on the life of Michelangelo de Buonarroti: The Agony and the Ecstasy, Signet Books, 1961

"[...] yet, the craftmanship of nature provides extraordinary pleasure for those who can recognise causes in things and who are naturally inclined to phylosophy.
        Aristotle, in 'de partibus Animalium', quoted by Johnson L., 1981, The thermodynamic origin of ecosystems, Can. J. of Fisheries and Aquatic Sciences, 38: 571-590

TABLE OF CONTENTS Page

Section 1

CATTLE, STRAW AND SYSTEM CONTROL
AN INTRODUCTION

Chapter 1

CATTLE, STRAW AND SYSTEM CONTROL:
AN INTRODUCTION

Fibrous crop residues are a major feed resource in many farming systems. However, not everywhere they are considered valuable, and if they are useful, still they can be fed in many ways. Therefore, this thesis addresses two questions regarding the use of straws as ruminant feed. In the first place it attempts to specify the suitability of straw feeding methods in mixed farming systems. Secondly, and on a more abstract level, the thesis tries to describe the role of straw in the shape and drive of farm systems, indirectly providing background on the answer to the first question. Since it is impossible to describe the use of straw in every conceivable system, it is necessary to identify a few major cases at strategic points in the development of farming systems to achieve general understanding, rather than an *ad hoc* description. This thesis attempts, therefore, to answer the two questions by placing the systems on scales that represent feed quality and quantity, access to inputs, as well as types and levels of animal production. The scales were constructed by studying systems on a range that varies between high and low input use in temperate and tropical areas, in both ancient and modern history. A brief review of role and type of animals and feeds of various farming systems is given before the origin, structure and objectives of this thesis are discussed. This is necessary because animals are the means through which the energy from straws and other feeds are converted into outputs that play a role in the organisation, i.e. control of societies, whether on farm, at a regional or a national level.

LIVESTOCK AND FEED BIOMASS

In many societies livestock represent wealth and / or power (Box 1). The etymological links between animals, capital, money and savings in unrelated languages (Renfrew, 1994) was already observed in the Roman empire. Columella says in the first century AD:

> *'[...] grazing is the oldest and most profitable form of agriculture, a reason why the words for money and capital are apparently derived from words for livestock. This is because our ancestors have had this, and because in many societies this form of wealth is still existent'* (translated from Ahrens, 1976).

These linguistics can be explained by seeing animals as transformers of solar energy from plant biomass into products useful for society (see for example Odum, 1971). In practical terms, such livestock products include dung, fibre, food, pleasure, security, speed, status or work. The relative importance of these products depends on what this thesis calls resource / demand patterns, defined as:

> *the pattern of relative access of farmers and farm communities to land, labour, energy from feed biomass or fossil sources on the one hand, and prevailing demand for (animal) produce on the other.*

Box 1: LINGUISTIC LINKS BETWEEN WORDS FOR CATTLE AND WEALTH

- ***Cho-Chiku*** (Japanese: saving money) consists of two characters, of which the first *Cho* means saving. The second word is also used for livestock though the character is (only partly) different, *Chiku*. The Chinese etymology is very much similar,
- ***Råjåkåyå*** in Javanese literally means *rich king*, but it has the meaning of wealth and cattle. ***Rejeki*** is closely associated with the Javanese *råjåkåyå*, but equally so with the Farsi and Urdu word *reejek* which stand for wealth. Whether reejek relates to animals or cattle could not be traced.
- ***Ente*** means cattle in Lunyomkole (a Bantu language from Uganda), and *sente* means money in that same language,
- ***Mikne*** (Hebrew) for cows, goats, camels, etc.. It consists of the rootword *kne* or *kana*, that means to buy, and an affix *mi* that makes the root into a noun.
- ***Byoto*** (Polish) means cattle and originates from a slavic root word *bydło* which relates to the meanings of 'being, standing, living, the house, possession'. This root meaning still survives in Czech and Slovakian but it has disappeared in Polish. The change of meaning from possession to livestock is typical for many Slavic languages.
- ***Da*** (Celtic/Welsh) means goodness or value as well as head (number) of cattle. In the same language, ***gwerth*** means value or worth, linked with *gwartheg* that means cattle.
- ***Vee*** (Dutch), ***Vieh*** (German) for *livestock* is related to *fee* (English) and originates from *fehu* (Old Saksish) which means both livestock and wealth or money. Compare the *fia* (Old frisian), *faihu* (Gothic), *fe* (Norwegian) en *fä* (Swedish).
- ***Cattle*** relates with capital via *caput* (Latin: head, number of e.g. animals); the word ***chattle*** seems to be an intermediate.
- ***Ganado*** (Spanish: livestock) is related to *ganar* (Spanish: to earn, to win, to gain).
- ***Pecunia*** (Latin: wealth, money) is linked with *pecu* (livestock) and also used in the Spanish word for animal husbandry (*pecuaria*).

sources: Slicher van Bath (1963), De Vries (1973), Webster (1984); Longman (1985), pers. comm. from Arieli (1991), Poedjono Sardjono (1991), Dana Subrtova (1992), Grazena (1992), Sedrace (1992), Yoichi Matsuki (1992), Hu Zhihong (1993), Hugh Jones and Patricia Martin (1994) and pers. obs.

Within and between regions and communities a large variation of farming systems is possible, but livestock is traditionally an asset, particularly for farmers that have access to abundant land and - by implication - feed biomass and solar energy. A higher demand for crops makes, however, that common or private grazing lands become scarce (Jodha, 1986; Alexandratos, 1988; Kelley and Rao, 1994). This process started long ago, and continues till today in many parts of the world. The Russian agricultural economist Chayanov observed for his country some 70 years ago:

> *'[...] it is by no means everywhere that peasant farms can have such an abundance of meadow. In the North of Russia, only in Vologda guberniya are there settlements where there is more than a desyatina of meadow to each desyatina of arable[1]. In other places, due to the pressure on the land, meadows have to be plowed and feed getting transferred to the arable [...]'* (Chayanov 1926, p.160 in Thorner et al., 1966).

The farmers' response to changed access to feed differs between systems. Initially, *'feed getting from the arable'* may be done by the production of fodders, tubers and catchcrops[2], e.g. in the Flemish and Norfolk rotations in north western Europe of centuries ago (Lord Ernle, 1961; Slicher van Bath, 1963). Such systems still exist today on, for example, the riverine soils of the Gangetic plains in India, Pakistan or in the Nile delta of Egypt, using legumes and cruciferae as catch and fodder crops. With a higher demand for crops than for animal produce, however, the feed from crop residues becomes the major source of energy for livestock.

FEED AND CROP RESIDUES

The term crop residues is a name for a variety of feeds that result from the production of crops. In the definition of this thesis they consist of fibrous feeds such as straws and byproducts from oil seed pressing and grain milling. At a more abstract level, this thesis stretches the definition to include rejected grains and tubers or biomass from trees or grasses that are cultivated to protect or to support crops, e.g. in alley farming (Kang et al., 1990). Crop residues in this definition can therefore be of both high and low quality, a concept applied particularly in Ch. 5.2. The use of the term straw in this thesis indicates that the emphasis of the work is on the use of fibrous crop residues.

In some farming systems, straws can be valuable either for herd survival during feed shortage, or as a source of fibre in feeding systems where large amounts of concentrate and succulent feeds are used (Ch. 2.3). However, and in spite of its wide availability, the use of straw as animal feed has its limitations. The nutritive value, expressed in terms of digestible energy and protein content, is too low to support even low levels of milk and meat production if no other feeds are fed with it.

---

[1] gubernya is an administative unit; a desyatina is 1.1 ha.

[2] A catchcrop is a green manure that is planted primarily to "catch" soil nutrients that are mineralized in a period that the land would otherwise lie bare. The foliage and roots can either be fed to animals or incorporated in the soil.

Straws as feed for production of meat or milk

Several approaches are available to overcome the low nutritive value of straws for production of meat, milk or draught. They include:
- acceptance of temporary lower levels, or even loss of production (Ch. 5.1),
- supplementation of straws with better feeds (Ch. 3.1-3.4 and Ch. 4.1),
- improvement of nutritive quality particularly by chemical and/or physical or biological treatments (Sundstøl and Owen, 1984; Ch. 3.1-3.4).
- plantbreeding and management for more and better straw (Saleem, 1972; Reed *et al.*,1988; Joshi *et al.*, 1994).

It is of course also possible not to feed straw, but to use it for non-feed purposes such as for paper manufacture, fuel, roofing, mulch, production of chemicals and cultivation of mushrooms (Hartley *et al.*, 1987). The generally low nutritive quality and use of straws for non-feed purposes is well recognized since many centuries (Box 2).

Box 2:   THE USE OF STRAW FOR ON- AND OFF-FARM PURPOSES IN THE HISTORY OF THE MIDDLE EAST.

The use of straw for feed is mentioned in the bible as:
   'the lion shall eat straw like the bullock' (Isa 65:25; Isa 11:7).
However, the fact that straw is not primarily considered as feed, but more likely as bedding for the animals transpires in the words:
   'there is both straw and provender for our asses' (Judges 19:19)
and in
   'we have straw and provender enough, and room to lodge in'
or
   'straw and provender for the camels' (Gen 24:25 and 32).
The on-farm use of straw - not for feed - is also found in the mixing of straw with manure:
   'Moab shall be trodden down under him, even as straw is trodden down for the dunghill' (Isa 25:10),
and other off-farm use of straw for non-feed purposes is evident from:
   'Ye shall no more give the people straw to make brick, as heretofore, let them go and gather straw for themselves' (Exodus 5:7).

The principles of chemical and physical treatments were copied from the paper industry by workers such as Lehman and Beckmann around 1900. The first large scale practical application of this technology took place during and after the second world war in Norway (Homb 1984; Westgaard and Sundstøl, 1986). Work on NaOH treatment of straw in the tropics was done as early as the early forties in India (Kiran Singh, pers. comm., 1993).

Attempts at the improvement of straw quality with biological treatment is described already by Pringsheim and Lichtenstein (1920). However, this approach is unlikely to yield practical results, mainly due to practical difficulties of fermentation, organic matter loss, chances of toxicity and lack of proper micro organisms (Kiran Singh and Schiere, 1993). Biotechnological improvement of straws focuses mainly on the manipulation of rumen microflora, or on the use of enzymes on straws. Breakthroughs have not yet been achieved (Flegel, 1988; Hunter, 1991).

Plant breeding and management for improved straw quality and quantity appears to affect quantity of straw rather than quality. Environmental effects overshadow the genetic ones

(Joshi *et al.*, 1994). Broadly speaking this type of work, i.e. the choice of the proper crop, also includes the adjustment of cropping patterns vis-a-vis type of animals (Patil *et al.*, 1993).

The principle of treating straw with ammonia compounds such as urea were developed in the sixties and seventies, particularly in Asia. It became well established that urea treatment improves digestibility, intake and crude protein content (Jackson 1978; Jayasuriya and Perera, 1982; Perdok *et al.*, 1982; Saadullah *et al.*, 1982; Davis *et al.*, 1983; Ibrahim 1983; Sundstøl and Owen, 1984). Some ten years ago - in the early eighties - the remaining research issues on urea treatment of straw were:
- technical questions regarding the duration of treatment, the amount of urea and water to be used,
- the practical application of the method under field conditions,
- fundamental questions about the action of chemicals on the fibre, and of the intermediate metabolites on the animal metabolism,
- feasibility of straw feeding in different farming systems, e.g. issues of feeding calendars, economics of treatment and/or supplementation, and alternative uses of straw.

THE ORIGIN OF THIS THESIS

While working in Sri Lanka from 1983 till 1986, and from 1986 onwards mainly in India, I was associated with many of the above mentioned studies on treatment of straw, and their workers. The Straw Utilization Project (SUP), where I was employed, helped to solve some of the remaining technical and economic issues mentioned above (Ibrahim and Schiere, 1986; Schiere and Ibrahim, 1989). Much work on chemical and physical treatments was done in other parts of the world as well, reported for example in IAEA (1991), Ibrahim *et al.* (1992); Kiran Singh and Schiere (1993), Oosting (1993).

Besides the SUP work on aspects of practical application of urea treatment, the project also became involved in the question that finally became the main issue of this thesis:
what is the feasibility of straw feeding in different feeding systems, and can it be predicted without resorting to a large number of ad hoc trials?
In practical terms, the question was studied by doing economic calculations, on station and on farm trials, mostly reviewed in Ch. 4.1 and Schiere and Ibrahim (1989). Extrapolation of the results to a variety of conditions, however, necessitates a fundamental approach to understand the basic factors that determine the use of straw for feed or other purposes. While working at a deliberately abstract level, the second - and more intriguing - question of this thesis logically arose:
what drives, shapes and controls farming systems behaviour in relation to the use of crop residues for animal feed?

THE STRUCTURE OF THIS THESIS

To describe the role of straw as feed in a variety of farming systems, the central part of this thesis is divided over the sections: 2, 3, 4 and 5, each with individual sub. chapters. Section 2 describes[3] :
- the origin, the terminology, concepts and need of farming systems research in relation to straw feeding and livestock (Ch. 2.1),
- a classification of livestock systems relevant to the use of straw as feed or to other purposes, related with the causes and effects of technical innovations (Ch. 2.2),
- the methods of straw feeding in mixed crop-livestock systems, particularly in relation to changing access to grazing and straw for animal feed (Ch. 2.3).

Section 3 describes technical aspects of straw as animal feed. It compares the effect of urea ammonia treatment and different forms of supplementation on animal response parameters, focusing on questions like:
- which are the relative merits of urea ammonia treatment and/or supplementation with better feeds,
- which are responses to be expected in terms of feed intake, digestibility and animal response.

This section also provides parameters for the discussions about economics and feasibility of straw feeding technologies in Sections 4 and 5.

The question on the economics of straw feeding constitutes the key issue of the Section 4, divided over two sub-chapters, aiming to
- predict conditions where straw treatment might be an attractive technology for application by farmers (Ch. 4.1),
- test the nutritional validity of the coefficients and calculations of the previous chapter (Ch. 4.2).

The economic calculations of the previous section only consider the animals and the feed, essentially making feed supply dependant upon animal output. Since this thesis aims to understand the place of cattle and straw in the system on a more abstract and whole farm level, the fifth section aims to answer the following issues:
- can ration formulation concepts from temperate systems also apply to tropical systems (Ch. 5.1),
- how can animals and feeds be combined to achieve maximum system output, i.e. how do animal subsystems adjust to changes in quantity and quality of feed (Ch. 5.2). In other words, how does the feed supply shape the animal subsystem.

---

[3] The actual work sequence is that the study of Ch. 4.1 was done first while employed by the Straw Utilization Project (SUP) of Sri Lanka, under pressure of donors and local government to disseminate the technology of straw treatment with urea ammonia. It was followed by work such as reported in Ch. 3.1-3.4 to test the assumptions of Ch. 4.1. Work on duration, concentration of urea and amount of water for urea treatment are reported - with colleagues - in Ibrahim and Schiere (1986). The experiences with farmers and extension workers on practical issues of treatments is described in Schiere and Ibrahim (1989). After joining the Agricultural University in Wageningen for work in the Indo-Dutch BIOCON project, I found an opportunity for discussions, literature review and field observations about systems and modes of crop livestock integration (Ch. 2.1-2.3), testing of concepts and calculations of the Sri Lankan work (Ch. 4.2 and 5.1), as well as the final part (Ch. 5.2) and the Discussion (Ch. 6).

Finally, the discussion in Section 6 integrates the information and conclusions of the preceding chapters, while testing them against information and concepts from systems control, particularly from thermodynamic and information theory. The interaction between individual chapters is graphically represented in Fig 1.

Figure 1.   The structure of this thesis.

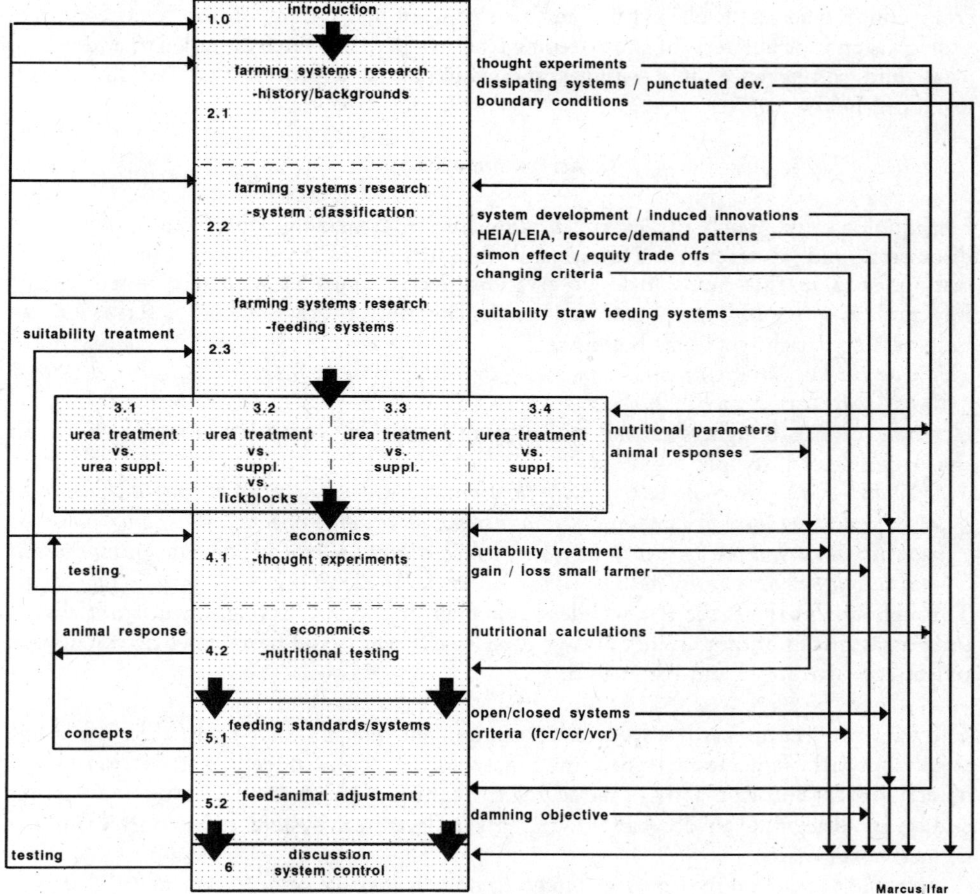

METHODOLOGY AND LAYOUT

The search for a comprehensive answer on the place of straw in farming systems, at an abstract level, arose from a mix of curiosity and exposure to a fascinating variety of systems and people. The systems ranged between irrigated and rainfed agriculture, subsistence and commercial animal production, near to and distant from cities in many parts of the temperate and tropical world. The people included farmers, extensionists, scientists from divergent disciplines, administrators and policy makers from local governments as well as from donor agencies. Each had his/her own concern and interest in the socioeconomic, ecological and social effects of straw feeding technologies. The discussions led to an exciting array and convergence of disciplines and concepts, in a process that will be briefly explained below.

Induction and deduction

Methodologically, much of Section 2 originates from reading, observation, reflection, discussion and abstraction. Testing and formulation of hypotheses can hardly be distinguished in this phase: field observations were compared with and tested against information from literature and vice versa. The final process is best described as an inductive approach to theory building:

if a particular (series of) phenomena occurs time after time, it must be possible to develop a theory that describes what is seen.

Generalization is fraught with the 'difficulty of exception', or what Lyklema (1991) calls the Archimedes principle:

[...] many events are composed of more than one sub-event, [...] what we might think to be the only driving force for the process is not necessarily the only one. [...]. By analogy with the buoyancy of, say, a piece of wood in water, which moves upward not because gravity would repel it, but because gravity pulls harder on water, we shall call this the <u>Archimedes principle</u>.

Throughout this thesis the Archimedes principle is found, e.g. development is not always progress, gains in output are not always gains in efficiency, and more cows do not always reflect access to more and better feed.

The work in Section 3 is less speculative, component research of the *'feed them and weigh them'* approach. It is meant to determine parameters for the modelling in Sections 4 and 5, sections that consist mainly of modelling and testing of the modelling results. It explores reality by designing an abstract model that reflects real system behaviour, mainly a deductive approach:

given the nutritional principles as known by now, it must be possible to predict the effect of changing resource / demand patterns on development of livestock systems.

Testing, terminology and analogy

Extensive testing of the ideas has been done throughout the drafting of this thesis. Theories and thoughts were subjected to unrelated literature and to a host of 'sparring partners'. The variety of disciplines and farming systems implies the use of terminology and situations that may not be clear to everyone. It was, therefore, necessary to include explanations in occasional footnotes. Extra literature references are included for those who are interested in further reading, not only to justify statements in the text.

The combination of literature, observations and discussions yielded the final output, mainly through a deliberate and extensive search for analogies. In this respect, special mention is to be made of a range of non-livestock-nutrition or now farming-systems literature that helped to grasp the 'feeling' of system behaviour. It is impossible to mention the specific contribution of each and they cannot be mentioned in references time and again, but the list includes, categorized per subject:
- the evolution, morphogenesis and function of systems (Darwin, 1859; Eisely, 1956; Dawkins, 1991)
- the progressive exploitation of systems to meet the effective demand of a society (Steinbeck, 1939; Garcia Marquez, 1977; Crosby, 1986; Rifkin, 1989, 1992; Ponting, 1991)
- the change of, and interaction between criteria for system success, i.e. the effect of resource / demand patterns on development in time and space (Harris, 1974, 1987; Crotty, 1980; Galbraith, 1987)
- the logic of trade offs between technical progress and socio-economic equity (Illich, 1974; Achterhuis, 1977, 1993; Ellul, 1989)
- the thermodynamic logic of system behaviour (Odum, 1971; Prigogine and Stengers, 1985; Gleick, 1987; Hawkins, 1988; Vroon, 1989; Dawkins, 1992; Lewin, 1993)
- the misleading distinction between 'science' and 'philosophy' (Gjertsen, 1989).
Each of these issues will surface throughout this thesis, and they will be integrated in the discussion.

DIFFERENCES, SIMILARITIES AND ABSTRACTIONS: CONCLUDING COMMENTS.

Differences are often a matter of magnitude rather than of substance. Differences may seem large, but that should not blind one for similarities. This problem was phrased effectively by a lady in a village near Bangalore (southern India) who asked me:
> *'you always say that farming systems differ from place to place and from time to time, but you also keep stressing their similarities, how can you do that ...?'*
She precisely expressed the tension between:
- <u>the need to distinguish systems</u> for effective targeting of research and extension,
- <u>the search for similarities</u> to extrapolate experience from one system to another.
An explanation from animal science could have been that from a nutritional point of view rodents, chickens, cows and even people can be put on one scale from small to large (Brody, 1945; Kleiber, 1961). In spite of their differences, these organisms are similar because they have four legs (wings are legs!) and because they require energy to reproduce and survive. Proper interpretation of similarities makes it possible, for example, to understand cow systems by studying mice. Depending on the objective of research, one focuses on either differences or on similarities. In the same way, one farming system can be understood by studying others, even though it would be foolish to use the blueprint of one for the development of the other.

Trends and mechanisms can be understood from analogies, but prediction does require verification. The abstractions in this thesis reduces the precision of prediction, but they allow an understanding of systems with their driving and shaping factors. The reply to the second question of this thesis ultimately depended on the use of principles from thermodynamics and information theory. Those disciplines provided concepts that I learned only after the start of the last article was made, and it provided unexpected support and backgrounds to the observations and loose ends of the previous chapters. Though tentatively, it provided a clue to discern a logical scale for development of farming systems. The reading of theory on behaviour of non-linear systems near and distant from equilibrium, the popularly called Chaos theory, filled the final gaps, at least as far as this thesis was concerned. It also opens a whole new area of hypothesis building and testing, left for other occasions.

ACKNOWLEDGEMENTS

Thanks are due to Ifar Subagiyo, who was the major sparring partner in the preparation of this chapter.

REFERENCES

Achterhuis, H., 1977. Filosofen van de Derde Wereld: Frantz Fanon, Che Guevara, Paulo Freire, Ivan Illich, Mao Tse-Toeng. Bilthoven, The Netherlands. 127 pp.

Achterhuis, H., 1993. Het rijk van de schaarste; van Thomas Hobbes tot Michel Foucault. Ambo, Baarn, the Netherlands. 365 pp.

Ahrens, K., 1976. Columella, Über Landwirtschaft, Ein Lehr- und Handbuch der gesamten Acker- und Viehwirtschaft aus dem 1. Jahrhundert u. Z., Aus dem Lateinischen übersetzt, 2., berichtigte Auflage 1976, Schriften zur Geschichte und Kultur der Antike 4, Akademie-Verlag, Berlin

Alexandratos, N., (ed.), 1988. World Agriculture toward 2000. An FAO study. FAO, Rome, Italy; Belhaven Press, London, UK. 338 pp.

Brody, S., 1945. Bioenergetics and Growth: With Special Reference to the Efficiency Complex in Domestic Animals. Reinhold, New York, USA. 1023 pp.

Crosby, A.W. 1986, Ecological Imperialism, the Biological Expansion of Europe (900-1900). Cambridge University Press. Cambridge, U.K. 368 pp.

Crotty, R., 1980. Cattle, Economics and Development. Commonwealth Agricultural Bureaux, printed by Unwin Brothers, The Gresham Press, Old Woking, Surrey, U.K. 253 pp.

Darwin, C., 1859. The Origin of Species by Means of Natural Selection. John Murray, UK. (revised version, 1985, Penguin Classics, UK. 477 pp.)

Davis, C.H., Preston, T.R., Haque, Mozammel and Saadullah, M., (eds.) 1983. Maximum Livestock Production from Minimum Land. Proc. of the fourth seminar, held from 2-4 May, 1983 at the Bangladesh Agricultural University, Mymensingh. Graduate School of Tropical Veterinary Science, James Cook University, Australia. 165 pp.

Dawkins, R., 1991. The blind Watchmaker, Penguin, London. 340 pp.

De Vries, J., 1973. Ethymologisch woordenboek, waar komen onze woorden vandaan?. Aulaboeken, nr. 6, Het Spectrum, Utrecht / Antwerpen.

Eisely, L. 1957. The Immense Journey; An imaginative naturalist explores the mysteries of man and nature. Vintage Books, USA. 211 pp.

Ellul, J., 1990. The technological bluff. Translated by G.W. Bromley. William B. Eerdmans Publishing Cy., Grand Rapids Michigan. 418 pp.

Flegel, T.W., 1988. An overview of biological treatment of crop residues. p. 41-65. In: Kiran Singh and Schiere, J.B. (eds.), 1988. Fibrous Crop Residues as Animal Feed. Aspects of treatment, feeding, nutrient valuation, research and extension. Proc. of an international workshop held on 27 and 28 October 1988 at the National Dairy Research Institute-Southern Regional Station, Bangalore; .publ. by ICAR, Krishi Bhavan, New Delhi, India; Dept. of Trop. Animal Production, Agricultural University Wageningen, The Netherlands. 269 pp.

Galbraith, J.K., 1987. Economics in perspective, a critical history. Houghton Mifflin Company, Boston, USA, 324 pp.

Garcia Marquez, G., 1977. Honderd jaar eenzaamheid (Cien Años de Soledad), Meulenhoff, Nederland. 428 pp.

Gjertsen, D., 1989. Science and Philosophy, Past and present. Penguin Books, Harmondsworth, Middlesex, UK. 296 pp.

Gleick, J., 1987. Chaos, making a new Science. An Abacus Book, London, U.K., 352 pp.

Harris, M., 1974. Cows, Pigs, Wars & Witches. The Riddles of Culture. Random House, New York.

Harris, M., 1987. Why Nothing Works. The Anthropology of daily life. A Touchstone book, New York. 208 pp.

Hartley, B.S., F.R.S., Broda, P.M.A. and Senior, P.J. (eds.), 1987. Technology in the 1990's: Utilization of Lignocellulosic Wastes. Philos. Trans. R. Soc. London, 321: 403-568.

Hawkins, S.W., 1988. A brief history of time, from the Big Bang to black holes. Bantam Books, Cox & Wyman, Ltd., Reading, Berks, UK. 212 pp.

Homb, T., 1984. Wet treatment with sodium hydroxide. In: Sundstøl, F., and Owen, E., (eds.), 1984. Straw and other fibrous by-products as feed. Developments in Animal and Veterinary Sciences, 14. Elsevier, 604 pp.

Hunter, A.G., (ed.), 1991. Biotechnology in livestock in Developing Countries. Proc. of an Intern. Conference on the Application of Biotechnology to Livestock in Developing Countries, held Sept. 4-8, 1989 at the Univ. of Edinburgh, Centre for Tropical Veterinary Medicine, Edinburgh, UK. 518 pp.

IAEA, 1991. Isotope and related techniques in animal production and health. Proceedings of a symposium, Vienna, 15-19 April 1991, Jointly Organized by IAEA and FAO. International Atomic Energy Agency, Vienna. 611 pp.

Ibrahim, M.N.M., 1983. Physical, chemical, physico-chemical and biological treatments of crop residues. p. 53-68. Pearce, G.R. (ed.), 1983. The utilization of fibrous agricultural residues. Australian Development Assistance Bureau, Research for Development Seminar Three, Los Banos, Philippines, 19-23 May, 1981. Australian Government Publishing Service, Canberra, Australia. 281 pp.

Ibrahim, M.N.M., and Schiere, J.B., (eds.) 1986. Rice straw and related feeds in ruminant rations. Proc. of an international workshop held from 24-28 March, 1986 in Kandy, Sri Lanka. Straw Utilization Project, Publ. No. 2. Dept. of Tropical Animal production, Agricultural University, Wageningen, the Netherlands. 408 pp.

Ibrahim, M.N.M., De Jong, R., Van Bruchem, J. and Purnomo, H., (eds.) 1992. Livestock and feed development in the tropics. Proc. of the International Seminar held on 21-25 October, 1991 at Brawijaya University, Malang, Indonesia. Agric. University, Wageningen, the Netherlands, Brawija University, Malang, Indonesia and Commission of European Communities, Brussels. 546 pp.

Illich, I., 1974. Energy and Equity. Calder & Boyars, London and Harper & Row, New York. 84 pp.

Jackson, M.G., 1977. Review article: The alkali treatment of straw. Anim. Feed Sci. Technology, 2: 105-30.

Jayasuriya, M.C.N. and Perera, H.G.D., 1982. Urea ammonia treatment of rice straw to improve its nutritive value for ruminants. Agric. Wastes, 4: 143-150.

Jodha, N.S., 1986. Common property resources and rural poor in dry regions of India. Economic and Political Weekly; XXI, No.27: 1169-1181.

Joshi, A.L., Doyle, P.T. and Oosting, S.J., (eds.), 1994. Variation in the quantity and quality of fibrous crop residues. Proceedings of a national seminar held at the Baif Development Research Foundation, Pune (Maharashtra-India), February 8-9, 1994. Syntec, Pune, India. 174 pp.

Kang, B.T., Reynolds, L., and Atta-Krah, A.N., 1990, Alley farming. Advances in Agronomy, 43: 316-359.

Kelley, T.G. and Rao, P.P., 1994. Availability and requirement of different sources of bovine feed in India. p. 87-97. In: Joshi, A.L., Doyle, P.T. and Oosting, S.J., (eds.), 1994. Variation in the quantity and quality of fibrous crop residues. Proceedings of a national seminar held at the Baif Development Research Foundation, Pune (Maharashtra-India), February 8-9, 1994. Published by the Indo-Dutch project on Bioconversion of Crop Residues. ICAR, New Delhi; Dept. of Trop. Anim. Production, AUW, Wageningen, The Netherlands; BAIF Development Research Foundation, Syntec, Pune, India. 174 pp.

Kiran Singh and Schiere, J.B., (eds.), 1993. Feeding of ruminants on fibrous crop residues. Aspects of treatment, feeding, nutrient evaluation, research and extension. Proc. of a workshop, 4-8 february 1991, NDRI-Karnal, ICAR, New Delhi, India. 486 pp.

Kleiber, M., 1961. The fire of life. An introduction to animal energetics. John Wiley and Sons, Inc., New York, London. 454 pp.

Lewin, R., 1993, Complexity. Life on the edge of chaos. Phoenix, London, U.K. 208 pp.

Longman, 1985. Longman Concise English Dictionary. Longman, Avon, U.K. 1651 pp.

Lord Ernle, 1961, English Farming, Past and Present, 6th edition, Heinemann, London / Melbourne / Toronto, Frank Cass & Co, London. 559 pp.

Lyklema, J., 1991. Fundamentals of Interface and colloid Science, Vol.1: Fundamentals, Academic Press, Great Yarmouth, Norfolk, UK.

Odum H.T., 1971. Environment, power and society. Wiley Interscience, New York, USA. 331 pp.

Oosting, S.J., 1993. Wheat straw as ruminant feed: effect of supplementation and ammonia treatment on voluntary intake and nutrient availability. Ph.D. Thesis. Agricultural Univ., Wageningen, The Netherlands. 232 pp.

Patil, B.R., Rangnekar, D.V. and Schiere J.B., 1993. Modelling of crop-livestock integration: effect of choice of animals on cropping pattern. p.336-343. In: Kiran Singh and Schiere, J.B. (eds.), 1993. Feeding of ruminants on fibrous crop residues. Aspects of treatment, feeding, nutrient evaluation, research and extension. Proc. of a workshop, 4-8 february 1991, NDRI-Karnal. ICAR, New Delhi, India. 486 pp.

Perdok, H.B., Thamotharam, M., Blom, J.J., Van Den Born, H. and Van Veluw, C., 1982. Practical experiences with urea-ensiled straw in Sri Lanka. In: Preston, T.R., Davis, C.H., Haque, M. and Saadullah, M. (eds.), 1982. Maximum Livestock Production from Minimum Land. Proc. 3rd Seminar, 13-18 February 1982, Bangladesh Agricultural Research Institute, Joydebpur, Bangladesh.

Ponting, C. 1991, A Green History of the World. Penguin Books, London. 432 pp.

Prigogine I., and Stengers I., 1985. Order out of Chaos, man's new dialogue with nature. Flamingo, London, U.K. 349 pp.

Pringsheim, H. and Lichtenstein, S., 1920. Versuche zum Anreicherung von Kraftstroh mit Pilzeiweiss. p. 29-39. In: Heuser, E. (ed.), 1920. Cellulosechemie, Jahrgang 1. Berlin.

Reed, J.D., Capper, B.S., and Neate, P.J.H. (Eds.), 1988. Plant breeding and the nutritive value of crop residues. Proceedings of a workshop held at ILCA, Addis Ababa, Ethiopia, 7-10 December, 1987. ILCA, Addis Ababa, Ethiopia. 334 pp.

Renfrew, C., 1994. World Linguistic Diversity. Scientific American, (January): 104-110.

Rifkin, J., 1989. Entropy into the greenhouse world. Bantam Books, New York, USA. revised edit., 354 pp.

Rifkin, J., 1992. Beyond beef: the rise and fall of the cattle culture. A Dutton book, Penguin, New York, 353 pp.

Saadullah, M., Haque, M. and Dolberg, F., 1982. Treated and untreated rice straw for growing cattle. Tropical Animal Production, 7: 20-25.

Saleem, A., 1973. A study of factors affecting the digestibility of paddy straw. MSc-thesis. G.B.Pant University, Pantnagar, UP, India.

Schiere, J.B. and Ibrahim, M.N.M., 1989. Feeding of urea-ammonia treated rice straw. A compilation of miscellaneous reports produced by the Straw Utilization Project (Sri Lanka). PUDOC, Wageningen, The Netherlands. 125 pp.

Slicher van Bath, B. H., 1963. The agrarian history of Western Europe, A.D. 500-1850. Translated by O. Ordish. Arnold, London, UK.

Steinbeck, J., 1939. Grapes of Wrath. A Bantam Book, New York, USA.

Sundstøl, F., and Owen, E., (eds.), 1984. Straw and other fibrous by-products as feed. Developments in Animal and Veterinary Sciences, 14. Elsevier, 604 pp.

Thorner, D., Kerblay B. and Smith R.E.F., 1966. A.V. Chayanov on the Theory of Peasant Economy. Published for the American Economic Association, by R.D. Irwin Inc., Homewood, Illinois. 317 pp.

Vroon, P.A., 1989. Tranen van de krokodil, over de te snelle evolutie van onze hersenen. Ambo, Baarn, The Netherlands.

Webster, 1984. Webster's Ninth New Collegiate Dictionary. Published by Merriam Webster Inc.

Westgaard, P., and Sundstøl, F., 1986. History of straw treatment in Europe. pp. 155-163. In: Ibrahim, M.N.M. and Schiere, J.B. (eds.), 1986. Rice Straw and related feeds in ruminant rations: Proceedings of an international workshop, Kandy, Sri Lanka, SUP Publication No.2. 407 pp.

Section 2

FARMING SYSTEMS RESEARCH:
CONCEPTS, CLASSIFICATIONS
AND INDUCED INNOVATIONS

Ahí viene - alcanzó a explicar - un asunto espantoso como una cocina arrastrando un pueblo. En ese momento la población fue estremecida por un silbato de resonancias pavorosas y una descomunal respiración acezante. [...] Pero cuando se restablecieron del desconcierto de los silbatazos y resoplidos, todos los habitantes se echaron a la calle y vieron a Aureliano Triste saludando con la mano desde la locomotora, y vieron hechizados el tren adornado de flores que por primera vez llegaba con ocho meses de retraso. El inocente tren amarillo que tantas incertitumbres y evidencias, y tantos halagos y desventuras, y tantos cambios, calamidades y nostalgias había de llevar a Macondo.
    Gabriel Garcia Marquez, 1984, Cien años de Soledad, septima edición, ESPASA-CALPE S.A., Madrid

De boer tracht de krachten van de natuur en de menselijke arbeid onder gebruik van hulpstoffen te combineren om zodoende goederen te produceren welke direct of indirect nodig zijn voor zijn levensonderhoud. [...] Oorspronkelijk was hij daarbij grotendeels afhankelijk van de factor natuur die hem haar wetten stelde. Ook de noodzakelijke arbeidskracht was door het ontbreken van technische hulpmiddelen beperkt en weinig gedifferentieerd. Langzamerhand heeft hij zich echter door zijn vindingrijkheid enigszins aan de willekeur van de elementen kunnen onttrekken dan wel de invloed van de wetten kunnen wijzigen naar zijn wil. In enkele takken van het landbouwbedrijf is de productie welhaast tot een beweging "los van de natuur" geworden [...]
    Vondeling, A., 1948, De bedrijfsvergelijking in de landbouw, proefschrift Wageningen

Chapter 2.1

LIVESTOCK AND FARMING SYSTEMS RESEARCH
(I): HISTORY, CONCEPTS AND FUTURE[1]

J.B. Schiere

SUMMARY

The role of straw in animal feeding systems depends on: a) the functions of straws and animals in agriculture, and b) differences between systems which use straw to keep animals. The use of the Farming Systems Research (FSR) methodology as a framework for such a discussion is however complicated by the confusion about its concepts. Even the words farming and systems can be interpreted differently. A review of ancient and modern FSR in this chapter explains several forms of this type of work, the use of 'thought experiments' for the design of new farming systems, as well as differences and similarities between crop- and livestock systems research (CSR and LSR). It is argued that in spite of its long history, FSR continues to be necessary because systems are constantly changing and agriculture remains important in many societies. The use of straw for feed - on the interface between crops and animals - is an archetypal topic for a systems approach, particularly in low input conditions where limited access to resources forces crop and livestock sub-systems to interact more. The practical impossibility of testing each innovation under variable conditions, requires understanding and identification of the driving factors behind system behaviour.

---

[1] This Chapter is based on a paper "Livestock Systems Research in the Tropics, a Review of Dutch Experiences", presented at the Global Workshop on Animal Production Systems, 16-21 September 1991 at San José, Costa Rica

INTRODUCTION

Much research has been done on technical aspects of feeding of straw to livestock (Sundstøl and Owen, 1984; Kiran Singh and Schiere, 1993). Whether and how the various types of straw should be used for animal feed has however, no single answer because:
- the use of straw as feed depends on the type, and cost of straw, the role of animals in the system, the access to other feeds, and the use of straw for other purposes,
- differences between systems imply that what is useful in one place, may not be appropriate for another.

A review of the functions of straw and animals, and a classification of systems, is therefore needed to evaluate the role of straw as a feed at more than a site specific level. The study of interactions between crops and livestock via the use of crop residues as feed also typically requires a systems approach. The question then arises about the definitions of systems research and the possibility to use existing terminology and concepts.

An apparently logical basis for a study of systems, is the use of concepts from what is often called Farming Systems Research (FSR). Unfortunately, FSR suffers from a confusing array of definitions, methodologies and objectives, only partly overcome by attempts at formalization (Shaner *et al.*, 1982; Fresco, 1986; Merril Sands, 1986; Simmonds, 1986; Fresco and Westphal, 1988). It is therefore necessary to review this terminology and its concepts, here done by looking at ancient and recent FSR, and by giving a broad interpretation to the term FSR. The review has a bias towards Dutch work, but it illustrates the issues at stake. More information can be found in publications that have provided a basis for many of the ideas expressed below, which reflect a mix of agronomic-, livestock-, economic- and historic- disciplines (Trow-Smith, 1957, 1959; Slicher van Bath, 1963; Thorner *et al.*, 1966; Nou, 1967; Crotty, 1980; Ruthenberg, 1980; Fresco, 1986; Ponting, 1991).

This chapter is the first of a series of three, that form the introduction to a study on the use of straw as cattle feed. Besides reviewing the concepts and objectives of ancient and recent FSR, it discusses differences between cropping systems and livestock systems research (CSR and LSR), as well as the rationale for further FSR. The review serves to categorize the forms of FSR relevant for this study, and to identify existing terminology. The second chapter proposes a classification of livestock systems related to the importance of straw feeding. The changing role of animals and the use of crop residues in mixed crop-livestock systems is discussed in the third chapter.

FARMING SYSTEMS RESEARCH: TERMINOLOGY AND CONCEPTS

The terms 'system' and 'farming' need clarification before discussing the composite and broader term 'farming systems research'. The word 'research' will not be elaborated, but it is important to state here that - in our concept - research can be done by people ranging from farmers via clergy and scientists to administrators, as shown in this chapter.

Systems

The word system has several meanings (Longman, 1985). The relevant ones here are a) **unit** and b) **method** of operation. They may be two sides of one coin, but their distinction helps to avoid confusion.

A system in the sense of a **unit** can be defined as:
> *an arrangement of components or parts (e.g. subsystems such as animal, farm, national economy) that interact according to some process and transform inputs into outputs.* (Odum, 1983; Fresco and Westphal, 1988).

FSR that refers to systems as the unit of operation is reflected in the definition of Shaner *et al.* (1982) for farming systems research and development (FSR&D):
> *an approach that focuses on the household as the core activity, that manages the other productive activities.*

The unit approach is also followed by Hart (1982) and Norman and Gilbert (1982). Fresco and Westphal (1988) use the term **farm system** to refer to this unit of farming:
> *a decision-making unit comprising the farm household, cropping and livestock systems, that transforms land, capital (external inputs) and labour (including genetic resources and knowledge) into useful products that can be consumed or sold.*

A system in the sense of a 'mode', or 'method' of operation can be defined as:
> *an organized or established procedure'* (Longman, 1985).

The mode or method of farming, is implied in the term **farming systems** as used by Ruthenberg (1980), **feeding systems** of Ch. 5.1 or the 'bedrijfsstijlen' (= **style of farming**) in the sociological work of for example Van Der Ploeg (1994).

Farming

Farming, such as agriculture, includes activities such as cropping, animal husbandry, fisheries, forestry, and horticulture. FSR ideally considers the farm(ing) systems in relation to off-farm activities and consumption. However, the need for limitation and simplification forces systems' researchers to impose boundaries, a case of reductionism that is contrary to the holistic view that FSR strives for. Even Shaner *et al.* (1982) focuses on cropping systems research (CSR), hardly mentioning livestock systems research (LSR), and excluding non agricultural activities. Some recent work intends to overcome this disciplinary focus by looking at more components of the system (Stroosnijder en Van Rheenen, 1993).

The very need to establish system boundaries, harbours a danger and a difficulty. The danger is that productivity of land or livestock system is considered without taking into account the side-effects in other systems, the so-called externalities. This can give a false sense of achievement and sustainability in a subsystem (Conway and Barbier 1990; Daly and Cobb 1990).

The difficulty of boundary setting is that one does not know where to stop, neither in time nor space. Upon deciding that the animal or crop level is too limited as unit, one can proceed to farm, village or regional level (Hart, 1982; Shaner *et al.*, 1982; Fresco, 1986; Figure 3.2). System aggregation is carried very far by workers who see the world as one system (Hopkins and Wallerstein, 1982). It is also expressed in the Spaceship Earth concept

(Buckminster Fuller, 1975) more rigidly described by Cooke (1974) and Odum (1983). The GAIA theory of Lovelock (1979) even considers the earth as a self-regulating organism, a concept that is quite possible within the definition of a system as unit! Importantly, their organicism does not imply that animate and inanimate, individual and conglomerate systems are the same. However, it recognizes that in some respects, both can behave in a similar manner: as an organism. In ecological terms such systems are called dissipative structures, a concept that is further discussed in the final part of this thesis (Ch. 6).

Forms and objectives of FSR

The term FSR is here used to imply the aggregate of a variety of interdisciplinary and holistic studies of farm(ing) systems, named FSR *sensu latu* by Simmonds (1986). Many 'modern' FSR concepts such as surveys, on farm research, and farmers' participation, are not new. They were part and parcel of development projects and systems research, long before the formalization of FSR in the last two decades. The recent formalization of terminology and approaches has helped to provide a logical sequence between the different stages of FSR. The classification of Table 1 resembles the categorization of FSR by Simmonds (1986). It intends to explain, but not to imply sharp boundaries between stages of FSR (Box 1), each one of them obtaining value as a part of the entire sphere of work.

Table 1. A categorization of stages within FSR *sensu latu*

FSR *sensu strictu*
A.   compilation of basic system information and data concerning field situations

B.   development of system concepts, research methodologies and software

FSR&D or FSR&E
C.   1) farming systems description using techniques such as surveys, rapid rural appraisals, agro-ecozoning

C.   2) identification of innovations, design of relevant (component) research

C.   2/3) modelling or design of farming systems, (farm synthesis, NFSD, ex-ante and ex-post analysis)

C.   3) on station technology testing,

C.   4) on farm technology testing, with an emphasis on farmers participation

C.   5) demonstration and extension of results, monitoring

component and upstream research
D.   animal science, crop science, agronomy, soil science, sociology, anthropology

Note: see explanation in the text.

FSR - sensu strictu

Simmonds (1986) would call stage A and B of FSR in Table 1 the:
> *'FSR-sensu strictu: the deep analysis of farming systems as they exist, essentially an academic activity'.*

It provides indispensable information for C1-5 (FSR&D or FSR&E), together with disciplinary information from stage 'D'. The A and B stages also cover what is called Farming Systems Analysis by Merril Sands (1986), and the work in systems studies by authors such as Spedding (1988) and De Wit (1992). The search in this thesis for driving factors behind system change, as well as for criteria of system success and classification (Ch. 2.2-2.3 and Ch. 6) belongs in stage's A and B. Being a more academic approach than the practically oriented forms of FSR, this form can more explicitly employ terminology and concepts from sciences such as ecology, economics and thermodynamics. In essence however, it aims to understand systems in an inductive manner. From a large number of observations, it tries to develop laws with general validity.

FSR&D or FSR&E

The C1-5 stages reflect the steps in applied systems work, variations on the theme of Shaner *et al.* (1982). In Simmonds' terms they are:
> *'OFR/FSP (on-farm research with a farming systems perspective), a practical adjunct to research which seeks to test the socio-economic suitability of research ideas on farm before recommending extension'.*

The distinction between the Farming Systems Research and Development (FSR&D) as used by Shaner *et al.* (1982), and Farming Systems Research and Extension (FSR&E) as employed by Amir and Knipscheer (1989) is not essential here (Hildebrand and Waugh, 1986). The inclusion of modelling or design of farming systems (stage C2/3: modelling or design of farming systems) into FRS/D illustrates that formal classifications cannot cover all concepts. It is also meant to imply that C2/3 is an integral, rather than an independent component of FSR&D.

NFSD

The C2/3 form of FSR includes what Simmonds calls:
> *New Farming Systems Development (NFSD), which in its extreme form seeks to develop complex radical changes rather then the stepwise change characteristics of OFR/FSP'.*

He refers mainly to the design of new systems such as alley cropping (Sumberg and Atta-Krah, 1988; Kang *et al*, 1990). Specific examples of such NFSD in livestock development are the design of new ley farming systems (Gibson, 1987), pig-fish-crop systems (NRC, 1981), or new dairy systems (Biewinga *et al.*, 1992). In a sense, the NFSD is a form of *ex ante* analysis, i.e. modelling that serves to understand systems as done in Ch. 4.1 and Ch. 5.2. It is based on existing information, to generate and test new ideas or system shapes, i.e. it has a more deductive approach. NFSD can be done at the farm, but also at regional or higher levels e.g. Veeneklaas *et al.* (1991) and Struif Bontkes (1992). However useful the distinction between complex radical and stepwise change may be, it can only be drawn on pragmatic grounds, not as a matter of principle (Box 1). Basically, both FSR&D and NFSD aim to affect system changes.

Thought experiments

An important form of FSR in this thesis, and here considered to be a part of NFSD, resembles Von Thünen's *'Gedanken Experimente'* (**thought experiments**). Thought experiments are also known in other disciplines (Dawkins, 1982). Here they employ a variety of calculations to understand existing or to design new farming systems, often assuming simplified conditions to formulate general laws (Nou, 1967 and De Wit 1968).

Box 1:   ABRUPT *VERSUS* GRADUAL TRANSITION BETWEEN SYSTEMS

The reasoning in this thesis mostly assumes a gradual change between systems, as expressed by the fragment form M.C. Escher's 'Metamorphose III'. There are however indications that systems undergo abrupt changes as they move in time and space. Two kinds of abrupt change seem to occur: here to be called orbit change and punctuated development.
- orbit change is caused by indivisibility of production factors, just like indivisibility of quantums causes electrons not to linger between orbits
- punctuated development relates to the fact that some systems tend to be very unstable at certain points of time / space. It is tempting here to think in terms of **attractors**, a concept from chaos theory, described for evolution and physics by Prigogine and Stengers (1985), Gleick (1987), Vroon (1989), Dawkins (1992) and Lewin (1993). Probably, the issue can also be related to the entropy watersheds of Rifkin (1989).

The introduction of a large animal or a piece of equipment into a small farm system, is likely to cause orbit changes, due to indivisibility. But even seemingly gradual changes such as the use of fertilizer can cause a farm system to undergo punctuated development: harvesting dates, feed conservation practices, labour requirements and feed purchases need to be modified, resulting in a new farm design. The concept of agricultural involution (Geertz, 1963) implies that refinement i.e. gradual change is possible within a given framework. The earlier mentioned use of NFSD for either abrupt or gradual change might be based on choice or orientation. It might however also be forced upon the researcher / farmer by inherent system dynamics due to 'chaotic behaviour', a so-called breakthrough.

Von Thünen used them to determine the boundaries of his Standörte (**locations**), an early form of systems classification. The quest for simplifications was expressed by the saying *'let* *us assume...*'[2]. Von Thünen's predecessors such as Young used thought experiments to determine optimum farm size (Nou, 1967). Modern LSR in the Netherlands uses them to design new farming systems (Tiesema, 1980; Keuning, 1987; Biewinga *et al.*, 1992). The view of Merril Sands (1986), that NFSD excludes farmers' participation does injustice to the prominent role of farmers, also in this activity. Farmers' participation in NFSD is by no means impossible, rather it is an almost common activity in many systems, see for example Elogoet and Van Gils (1989) and the work of the Friesian farmer P.B. de Boer. The latter pioneered the use of nitrogen fertilizer on dairy farms, sometimes in the face of the scepticism of extension workers and fellow farmers (Bakker, 1973; Anon, 1974).

Thought experiments can use experience, intuition or computers, with a pencil on the back of a cigar box or with a stick in the sand of the farmyard. They are particularly useful where experimentation is difficult due to a large aggregation level of systems in time and space, or where high variation in farming systems precludes experiments. The results of thought experiments have to be verified, but they can save work by *ex ante* evaluation and hypothesis formation. The desk work and modelling for NFSD resembles the deductive approach of astronomy or physics that predicts the place of a phenomenon - a star or subatomic particle - before trying to detect it. The work in Ch. 5.2 is a typical case of such an approach, ideally a precursor to more practical forms of NFSD.

Component research

The inclusion in Table 1 of stage 'D', i.e. *component research* into FSR, stresses that the distinction between component and systems research is a matter of system hierarchy rather than of principle. Nutritional work at the animal level e.g. supplementation and treatment of straw rations in this thesis (Ch. 3.1-3.4) is systems work when seen from cell or tissue level, but it is component research when approached from a village or national level. Research on systems becomes FSR when it starts to tackle problems at the farm level, i.e. beyond the level of the individual animal or crop system. Partial budgeting studies farm problems but it is only a tool of FSR, whereas whole farm planning is FSR since it focuses on the system as an arrangement of components.

FSR IN THE PAST

The previous discussion clearly shows that FSR covers a broad range of activities. The following review of literature on FSR is not comprehensive, but serves to explain that many modern problems are not new. It further illustrates the variety of participants, objectives and approaches of FSR and it provides concepts that can still be used today.

---

[2] 'Let us assume...', the literal quote from Von Thunen is: 'Man denke sich eine grosse Stadt in der Mitte einer fruchtbaren Ebene gelegen, die von keinem schiffbaren Flusse oder Kanale durchströmt wird. Die Ebene selbst bestehe aus einem durchaus gleichen Boden, der überall der Kultur fähig ist. In grosser Entfernung von der Stadt endigt sich die Ebene in eine unkultivierte Wildniss, wodurch dieser Staat von der übrigen Welt gänzlich getrennt wird'. (quoted by De Wit, 1968).

Ancient FSR

FSR in its most basic form has been a precondition for farmers' survival since the beginning of agriculture. Farmers' participation in that form of FSR is assured. Some individuals however - such as priests and rulers - may accumulate more and other information than the cultivators or animal-keepers themselves. In fact, the first written studies of farming systems are known from about 3000 BC! Administrative records kept by temples of city states in Sumerian society, tell the story of changes and problems in their agricultural system (Ponting, 1991). These early records are likely to be written by and for administrators. Farmers' participation was probably remote and the information was likely to differ from that collected by farmers themselves.

Various forms of FSR were done in the Roman empire by authors such as Cato and Columella (Rose, 1954; White, 1970), though some of that had poetic rather than historic relevance. Interesting reviews of livestock diseases in the ancient systems are given by Blaisdell (1994) and Bodson (1994). A more recent example of FSR is reported in the Domesday book of 1086 AD. That book contains detailed descriptions of British farming systems in those days (Trow-Smith, 1957; Lord Ernle, 1961). In its scope it resembles the work by the Portuguese and the Spanish settlers in South America (Slicher van Bath, 1979) and by colonizers elsewhere in the world who practiced FSR to collect data for administrative purposes or for academic interest. Randhawa (1982) in his history of agriculture in India - a form of FSR itself - describes FSR by early colonists in India, the Islamic conquerors. He states that:

> 'one Ibn Khurdaba, a high official of the Caliphs of Baghdad, who died in AD 912, employed his leisure in [..] researches, resulting in his book Kitàb-l Masàlik wà-l Mamàlik (Book of Roads and Kingdoms) in which he provides an excellent description of early Sind (Indian) people and agriculture'.

Ancient work, recurring issues

The early FSR practitioners share agricultural problems and methodological FSR issues with modern society. Their mention of problems concerning sustainability, salinization, deforestation, and resulting ecological as well as socio-political instability should provide clues for modern policy makers (Lockeretz, 1978; Crosby, 1986; Ponting, 1991). Recurring issues are also shown in relation to the use of straw and to the feeding of animals. Straw had multiple uses even in Pharaonic times, e.g. it was used for brickmaking in Egypt (Exodus V). In Roman times, Columella refers to the need to adjust the animal to the feed (White, 1970), a comment still valid in low input farming of today, and a central issue of this thesis (Ch. 5.1 and Ch. 5.2). Nou (1967) tells how both Young and Thaer attempted to establish the ideal farm size, i.e. the 'proportioned farm' and 'rational farming'. Obviously, such a static concept does not exist in changing conditions, but the principle may be valid and has been traced back to the Romans (Nou, 1967).

Methodologically, FSR tends to collect too many data. The need for rapid rural appraisals was thus felt already 900 years ago in the preparation of the Domesday book:

> 'the information amassed by the commissioners was, in fact, so full that it became unwieldy. The inquisition in the rest of England was continued with a restricted questionnaire.' (Trow-Smith, 1957, quoting J.H. Round),

Collection of - reliable - data has never been easy either. The name 'domesday' seems to come from doom that was preached for those that did not give proper information for the ruler's administrative purposes (B. Slicher van Bath, pers. comm. 1992). The problem of obtaining reliable information is beautifully expressed by Backer (1934) for his flora of weeds in Java. He characterizes his list of plant names as:
'*having been recorded by a fool from the mouth of a story teller*'.[3]

FSR in the last two centuries.

FSR of the last few hundred years seems to differ from more ancient forms in the greater emphasis on experimentation in the former. If this is true, a point yet to be proven, it is a change consistent with developments in other sciences, i.e. in the reductionist and mechanistic footsteps of Bacon and Newton (Dijksterhuis, 1975; Achterhuis, 1990). Swanson and Claar (1984) traced the beginnings of agricultural education to the early 17th century, stating that it was inspired by workers such as Rabelais (AD 1483-1553) and ultimately Romans such as Cato and Columella! Lord Ernle (1961) reviewed early farm literature from the early 16th century, and also related it to the work of the Romans!

The reviews of Thorner *et al.* (1966), Nou (1967) and Hayami and Ruttan (1985) explain a tradition developed by great but often forgotten agronomists and economists of the last few centuries, e.g. Young, Bakewell, Von Liebig, Von Wulffen and Chayanov. The place and tradition of FSR by the Dutch geologist/agronomist Staring in the last century becomes better understood by knowing his colleagues from abroad. Not insignificantly, Young, Von Wulffen and Staring developed some of their ideas while travelling, the first for leisure and curiosity (Lord Ernle, 1961; Nou, 1967), the second while in Napoleon's army (De Wit, 1969), and the third while on duty in the 'Ten Day Battle' of the Dutch against the Belgians (Veldink, 1970). Chayanov has travelled also, but he further witnessed many changes in a short span of time during the early decades of this century (Thorner *et al.*, 1966). The others, such as the Ibn Khurdaba quoted by Randhawa (1982), must have also had a good exposure to a variety of farming systems from extensive travel, allowing them to take distance and to see the wood for the trees.

Veldink (1970) describes how Staring liked to study the work of Thaer and contemporaries. Staring's ideas, that 'bedrijfsstelsels' (=farming systems) need to be distinguished by agro-ecological rather than administrative divisions, were possibly also influenced by Von Thünen. It must further be possible to trace Staring's influence to modern FSR and LSR in the Netherlands, where the practical application of component research is tested for different farming systems (Osinga, 1992). Equally so, the Dutch gentleman farmer P.B. De Boer (1907-1993) was a student of the German agronomists of the early 19th century. In

---

[3] The literal quote is: "Men zij gewaarschuwd tegen het stellen van te veel vertrouwen in de opgegeven inlandsche namen. De gebrekkige bekendheid van vele Europeanen met de op Java gesproken talen en met den plantengroei, de dikwerf hoogst gebrekkige plantenkennis der inlanders en hun zucht den informeerenden Europeaan tevreden te stellen of zich van hem af te maken, hebben samengewerkt om een reeks van waardelooze en onbetrouwbare plantennamen te scheppen, waarmede men boekdeelen zou kunnen vullen. De volksnamen, welke men vindt opgeteekend bij de planten onzer herbaria, zou men goeddeels kunnen definiëeren als namen, welke door een dwaas zijn opgeteekend uit den mond van een fantast."

his own way he has influenced the modern NFSD in the Netherlands (Anon, 1974; M. 't Hart, pers.comm., 1994).

Mention should also be made of the role of the clergy in FSR in tropical and temperate, colonized and colonizing systems. Many of them were, more closely than urban based intellectuals, involved in problems of agriculture. In the early part of the 19th century their curriculum in The Netherlands even included a compulsory course in 'landhuishoudkunde' (= agricultural economics). Examples of FSR by dutch clergy is found in the work of J.Kops, some 200 years ago (Veldink, 1970). De Vries (1994) mentions a preacher Alta who played a role in the 18th century fight against rinderpest. The glasshouse farmers of South Holland all know the name of 'pastoor Franciscus Verburch' (1616-1708), a Roman Catholic priest how allegedly introduced grapefarming there (Van Der Krogt, 1992).

Much recent systems work adds to, but still leans on schools of thought of the last centuries, as evident for example, in the work of De Wit (1969, 1992). For more information on FSR *sensu latu* the reader is referred to reviews with emphasis on economics in Europe by Nou (1967); to agronomy in colonial agriculture by Fresco (1986) and Scheltema (1926/27), and to a mix of historical and agronomic aspects by Slicher van Bath (1963), Campbell and Overton (1991), Bieleman (1992), Bech *et al.* (1980), Thorner et al. (1966), Lord Ernle (1961) and Cochrane (1979) for Europe, the Netherlands, Germany, Russia, and the US respectively. Randhawa (1980, 1982, 1983, 1986) reviewed aspects of ancient and recent Asian FSR.

MODERN TERMINOLOGY, OLD CONCEPTS

Whole farm approach and interdisciplinarity

Many of the 18th and 19th century FSR workers used a '**whole farm approach**', a concept natural to them, but lost to modern component researchers. The word holism is often used in this context, but it does not necessarily imply a mystical sense, i.e. where the *'whole is more than the sum of the parts'*. It is mainly used here in contrast with the reductionist approach of component research. As an example of a whole farm approach, it may serve to quote Von Wulffen, who wrote in 1823 that:
> *'the problem of soil fertility could be better understood by studying the input-output dynamics of the farm as a whole'* (quoted from Beets, 1990).

Von Wulffen, such as many other of these early workers was a scholar-farmer. They were agronomists as well as economists, a distinction that was made only later, and not without dispute about the need and danger for such reductionism (Nou, 1967). The relation between economics, agronomy and FSR was also recognised by Shaner *et al.* (1982) who said that:
> *'some argue that FSR&D is simply a modified version of farm management that has been widely practised during the 20th century [...]. While this claim has merit, the general feeling among those actively engaged in FSR&D is that FSR&D is new, at least as applied to the needs of small farmers in developing countries'.*

The topic of this thesis, i.e. that the use of crop by-products for the feeding of animals which in turn provide inputs such as dung or draught for crop production, is an archetypal problem for a whole farm and interdisciplinary approach. The work on identification of 'locations' where straw can be profitably fed, fits perfectly in the tradition set by Von Thünen's 'Standorte'.

Farmers' participation

Modern FSR aficionados who stress the need for farmers' participation can take heart from a statement by Baily in 1896 (quoted by Hayami and Ruttan 1985, p. 57):

> *'at the present time every intelligent farmer is an experimenter [...], this cumulative body of experience of the best farmers is capable of yielding better results than similar work which might be undertaken at an experiment station [...]. An experiment station, which is necessarily constituted for scientific research, cannot touch many of the most vital problems of farming'.*[4]

Young and his contemporaries were prime examples of scholar-farmers who led and initiated research. Young was an early student of NFSD and attempted to determine optimal farm size. Bakewell, was another such scholar-farmer and is considered the first to apply scientific principles to animal breeding (Fraser, 1949; Trow-Smith 1957; Lord Ernle 1961). All of these must have been preceded by unnamed farmers, who knowingly or unknowingly did the same type of work on a smaller scale. Anecdotal but interesting diaries of scholar-farmers show how their authors were interested in their farm as well as in their community, a holistic approach indeed. The author has had the privilege to some of the unpublished diaries of P.B. de Boer, the Dutch scholar farmer who led research on nitrogen use in dairy production in the Netherlands, before and after the Second World War. He was a worthy successor to the tradition of which Hellema was an example of the early 19th century (Algra, 1978) and Hemmema in the 16th century (Slicher van Bath, 1958). In terms of research, it was indeed a farmer who worked on the immunization of his cows against rinderpest some 200 years ago (Veldink, 1970), probably based on 'scientific' work of others (De Vries, 1994).

A review of farmers' participation and FSR is given by Farrington and Martin (1988). On the one hand, no development will take place if it is in the farmers' perceived interest. On the other hand, not all farmers have the same interests, i.e. the reference to farmers' interests is a hopeless generalization. Acceptance of one technology by a sector of the farmers can lead to the marginalization of many of their colleagues. Individual farmers' interests may clash with the requirements of society and even men and women may not agree on priorities (Olson, 1971; Bromley, 1992; Schiere, 1993). In terms of crop residue feeding, the interest of landless cattle owners may not be the same as those of land owning crop farmers (Ch. 4.1 and Ch. 5.2). Apart from the differences between farmers' versus urbanites' interests, the issue of clashing interests has also been shown in planning for sustainability by Posner and Gilbert (1991). Whereas many farmers look at the short-term and farm level, FSR scientists tend to focus on the long term and regional level.

---

[4] Hayami and Ruttan (1985) underline these points by saying that: 'even in nations with well developed agricultural experiment station systems, a significant portion of the total effort, until as late as the 1930 's or 1940 's, was devoted to the testing and refinements of farmers innovations and to the testing and adaptation of exotic crop varieties and animal species. It seems likely that even in the most advanced agricultural nations this activity contributed more to the growth of agricultural productivity than the more scientific work carried on by the experiment stations until at least the middle of this century'.

It is very clear however, that interactions between farmers and scientists - to name only two of the actors in agricultural development - are essential for development. Many of the questions raised in this thesis do in fact, originate from farmers, policy makers, researchers and the other actors in livestock development. They asked whether straw could not better be burned, used as mulch or for mushroom growing (Ch. 2.2 and Ch. 4.1). They also asked what difference treatment of straw would make financially and practically (Ch. 4.1). Farmers in particular raised the point that high biological yields are not always financially attractive (Ch. 5.1; Schiere, 1974). By doing so, both farmers and policymakers effectively directed this research to system issues that would not have been thought of otherwise.

LIVESTOCK SYSTEMS RESEARCH

Most of modern FSR focuses on cropping systems research (CSR). A review of 29 recent FSR projects in Indonesia, showed that 7 projects - including mixed farm work - could be considered to be LSR, with 22 projects on CSR (Marwar, 1989). However, a brief search of mainly Dutch literature and discussions with colleagues, uncovered a number of livestock projects and studies with FSR related components (Schiere, 1991). Well documented LSR was done by veterinarians in the colonial Dutch East Indies (Merkens, 1927; Aalfs, 1934; Hoekstra, 1948; Huitema, 1982). Their FSR is more 'farming system' than 'farm system' oriented, i.e. it focuses more on the method than on the unit of farming. It is preceded and parallelled by agronomist or economist colleagues in the Dutch colonies (Fresco, 1986), or by livestock colleagues elsewhere (Toulmin, 1984). Their experiences and motivations deserve separate study and they prove that even LSR is not a recent activity.

The need for a whole farm and interdisciplinary approach, i.e. the relation between livestock, crops and society, was well recognized by many of those colonial veterinarians. Hoekstra (1948) justifies his Ph.D. thesis by saying that
> 'in the first place it is important to determine the proper ratio between livestock, crops and forestry [...]'.

He also recognized the need for farming systems classification and multi-disciplinary approaches by stating:
> 'while describing the agro-ecozones of Indonesia, the agronomist needs to be advised by experienced ethnologists'.

Some of these officers had - or took more - time and personal interest for FSR than others. Their insight in systems was stimulated and facilitated by long field tours with many informal meetings, in the tropical sun, the monsoon rain or under lamplight on the porch of a village elder or a farmer (P. Hoekstra, pers. comm., 1993). This is in stark contrast indeed with the modern laptopped, linguistically handicapped and 'aeroplaned' consultant who needs Rapid Rural Appraisals to know what is going on.

CSR and LSR, how different?

Differences between LSR and CSR are commonly said to include aspects such as reviewed by Gryseels (1988):
- different duration and continuity of the production process,
- issues of multiple production goals,

- social/ psychological satisfaction/ attachment,
- problems of sample size, size of experimental unit and data collection / processing,
- mobility of animals,
- ownership patterns

Differentiation between CSR and LSR may be tempting, it is however too general and could be counterproductive.

Firstly, the points above are a generalization that implicitly compare work on annual grain crops with that on large ruminants. When the grain crop is replaced with a perennial (e.g. coconuts, apple trees), and the cow with a small animal (e.g. rabbit, goat or poultry), only the mobility issue remains a valid distinction between crop and livestock, if not the animals are tethered or stall fed. This multipurpose character is not unique for animals. The Sanskrit word 'Khamadhenu' implies that the cow is an animal fulfilling all men's desires, e.g. for milk, calves, urine or dung. But Sanskrit also has a similar word - 'Khalpavriksha' - for trees, implying that a tree can fulfil all needs e.g. timber, decoration, fuel, fibre, juice or brooms.

Secondly, it can be more fruitful for CSR and LSR to look for similarities or areas of shared interest, so that methodologies and concepts can be exchanged and enriched. More strongly, for those who work on mixed crop-livestock systems - as done in this thesis - it is even irrelevant to distinguish between LSR and CSR. Chayanov in Russia and his contemporary Aalfs in the Dutch East Indies for example, had an interest precisely in the interaction between livestock and cropping (Aalfs, 1934; Thorner *et al.*, 1966). Similarities and common interests between CSR and LSR are summarized in Box 2, showing that the distinction between CSR and LSR is therefore not a matter of principle, i.e. the general FSR terminology is also valid for the LSR in this thesis.

Box 2: SIMILARITIES BETWEEN CSR AND LSR

- the multipurpose function of livestock is also valid for, for example a coconut tree on a homestead that supplies, timber, shade, leaves for thatching and decoration, fibre, firewood, alcohol, spoons from the shell, brooms, or even 'milk, meat and oil',
- the problems of measuring feed intake and nutrient requirements in animals, are matched by those of soil scientists who have to predict the distribution, use and requirements of organic matter and plant nutrients in or on the soil.
- transport of feed for livestock in terms of location is matched in soil science by run off, run on, leaching, etc.
- the yield of grass or perennial crops and even annuals, is similarly affected by time and carry-over effects as is known from animal production. Mineralization of soil nutrients and compensatory gain in livestock cause similar problems.
- the problem of reliable data collection is similar for FSR, CSR, and LSR. The number and size of cattle can be kept secret by the farmer from a researcher, as well as the number and size of rice fields or coconut trees. This can either be caused by different perceptions of the term ownership (is the land mine if it is also owned by my father-in-law?), or by deliberate deception. The cattle herding tribes of the Fulani in West-Africa have ways to disperse their herds when the tax collector comes (Williamson and Payne 1965, p 418).
- plant, animal and all other systems have maintenance and production requirements (Odum 1971, Ch. 6).

REASONS TO CONTINUE FSR

Continued interest in FSR in all its diversity, epitomizes the relevance of the core concepts as well as the imprecision of the definitions. FSR can indeed serve as a convenient umbrella to justify a host of activities. The substantial amount of FSR up to date might make administrators doubt the need for further funding. It should also make researchers humble when pretending to embark on a new field of study. Where possible they should use previous insights and look for common ground with other disciplines. Why pursue FSR, and particularly this study on straw and livestock feeding?

The diversity of FSR requires a review of failures and successes in their different forms, rather than an evaluation of the conglomerate. This is an urgent task, but too large for this thesis. It is further complicated by the difficulty to test whether success or failure of any of these components is due to the conditions in which FSR was applied, or the approach itself. Also, the sequence in which the different stages were applied can affect the success of FSR. Moreover, little follow-up of the projects seems to take place and long-term adoption studies might show more positive results than often assumed (Tripp *et al.*, 1991). Both negative and positive but practical examples are presented here together with political and academic arguments to justify continued FSR for development and for the academic work of this thesis.

Negative and positive experiences with FSR

Many recent livestock development projects failed due to the lack of a systems approach (DGIS 1987; Gryseels, 1988). They often started for example, with the introduction of dairy cattle (FSR stage C5: demonstration and extension), without considering that animals can mean more or different things than only milk or meat. With the advantage of hindsight, it can be seen that insufficient attention was paid to the problem analysis (FSR stage C1-C3: FS description, identification of innovations, modelling or design, on station technology testing). Similarly, livestock production in industrial countries has ignored systems aspects, beyond household level, resulting in environmental and social problems. Durning and Brough (1991) as well as Rifkin (1992), in their criticism of the excesses of livestock production, could have made their point better, if they had distinguished between livestock systems (FSR stage C1: farming systems description). They do however signal problems that could have been prevented by more thinking at an earlier stage. Problems due to the lack of systems approach are - paradoxically - an argument in favour of FSR, provided it is well executed.

Positive examples of FSR can also be mentioned. In the first place, there is the work on agro-ecozoning (part of FSR stage C1: farming systems description) that distinguishes farming systems on the basis of agro-ecological and socioeconomic criteria. It permits a diversified development of agriculture, applied in the regional approach of Dutch livestock extension and research (Osinga, 1992), in the area of crop residue feeding also attempted in India (De Boer *et al.*, 1994). Secondly, there are good examples of the application of thought experiments (FSR stage C2/3: modelling or design of farming systems). They help to identify the farm systems where urea-treated straw might be useful (Ch. 4.1), and the *ex-ante* results are now confirmed more and more by field experience (Kumar *et al.*, 1993; Mahendra Singh *et al.* 1993). Paradoxically, in homogenous systems, the modelling part of

NFSD, thought experiments, is relevant, because the prediction of effects can be relatively accurate. In variable conditions, and in spite of lower accuracy, the thought experiments are even more essential. It is impossible in such conditions to do sufficient representative experiments for even major farm systems (Ch. 5.1). Promising NFSD is also done in the Netherlands (Biewinga *et al.*, 1992), to overcome problems which are beyond the individual farm system level. Thirdly, at the crop-livestock level, the increased attention paid to an holistic approach of the farm, has resulted in the recognition by plant breeders and economists of the value of straws for animals. Research is now done on the development of crops with dual value, for grain and livestock, a prime example of systems work (Nordblom and Halimek, 1982; Reed *et al.*, 1988; De Wit *et al.*, 1993; Joshi *et al.*, 1994).

Politics, change and variable farming systems

The remaining discussion on the reasons for continued FSR, is based on three quotes that refer to the political importance of agriculture, to a romantic notion about perfection in existing systems, and to the sustainability and complexity of systems with limited access to resources.

The political importance of agriculture was recognized by White (1970) for the Roman empire:
> '*the entire administrative structure of the Empire rested on the foundation of an agricultural surplus. Thus [..] a clear understanding of the agricultural methods and processes [...], is of utmost importance to the student of Roman history [...]. It has been asserted that what broke the back of the Roman empire in the West was the inability of the primary producer to maintain his vital role in the economy, in the face of continuing increases in taxation, low productivity and technical stagnation in agriculture*'.

The adjective Roman can be deleted here since both deficient and surplus agricultural production can weaken most if not all other systems (Odum, 1971; Ponting, 1991).

A romantic - and in our eyes incorrect - notion is expressed by Voelcker who studied traditional agriculture in India around 1889. He says that:
> '*in many parts there is little or nothing that can be improved... I, at least have never seen a more perfect picture of careful cultivation, combined with hard labour...*' (quoted by Carlier, 1987).

Such a statement may apply to static or slowly changing systems, but most if not all FS's are changing rapidly. (Ch. 2.2-2.3). Traditional knowledge systems are valuable (Chambers *et al.*, 1989; Reijntjes *et al.*, 1992), but rapid change makes it hard for farmers to keep abreast of the developments: research with a systems perspective is required as never before.

Sustainability, is an issue already known to the Sumerians, but rediscovered particularly since the publication of WCED (1987). It sets a series of goals for development, that necessitate system understanding where practical experimentation is impossible. Conway and Barbier (1990) say that:
> '*while it is relatively straightforward to attain one or two (...) goals, it becomes progressively difficult as more and more objectives are included in program and project designs. There are trade-offs, in terms of labour, time, skills, and capital, for the project and its staff, and for the farmers themselves. Choices have to be made - productivity at the expense of equity, for example, or sustainability at the expense of productivity.*'

Reduced access of low input farmers to resources, is likely to make processes and subsystems interact more strongly. Mixed farming occurs on a scale that ranges from systems with loosely connected diversified activities, to those with integrated but sensitive interactions between crops and livestock (Ch. 2.2-2.3). Because the definition of the system as a unit specifically mentions the relation between its components, integrated crop-livestock farming can be considered to be 'more system' than a diversified system. The increased and changing interdependency of crops and livestock require more rather than less systems work.

CONCLUDING REMARKS

This paper started by saying that the assessment of the role of straw for cattle feed requires a systems approach. Problems of using FSR concepts for such a discussion concern the confusion in terminology, the levels in system hierarchy and the variation in objectives of FSR. Even the word 'system' can be understood in at least two different ways: as a unit or as a method. Particularly the deeper sense of systems as a unit i.e. organism and sets of organisms will appear to be important for the reasoning and feeling of this thesis.

FSR in the broader sense, is an aggregate of activities, done for millennia and comprising a vast array of objectives, concepts and terminology. The review of recent and older terminology helps to categorize FSR and to put it in a tradition, avoiding confusion on semantics and the need to re-invent the wheel. Much LSR has been done in the past, often as an integral part of FSR. The distinction between CSR and LSR is vague and even irrelevant to many of its practitioners. Recent formalization of general FSR terminology and concepts, provides therefore a useful framework for the discussion in this thesis. The first three chapters of this thesis belong mainly to the category of FSR in the narrow sense. On-station testing/research of feeding systems is reported in Ch. 3.1-3.4, and on-farm trials were done, though not reported here. The *thought experiments* of NFSD are applied mainly in Ch. 4.1 and Ch. 5.2, chapters that help to explore system behaviour. In our context, the general differences between CSR and LSR are too small to introduce separate LSR terminology.

The continued need for systems work in livestock development is partly justified with the poor performance of projects without a systems approach, and by the positive results of systems work. The socio-political importance of agriculture also requires continued attention for the sustainability of agriculture. The use of crop by-products for animal feed, is an archetypal topic for FSR. It requires a multidisciplinary approach on an abstract level to predict the effects of change in a variation of systems that defy experimentation. The change of systems, the strong interaction between subsystems, and their multiple objectives complicate the prediction of results from innovations. It further stresses the need for a fundamental understanding of factors that shape and drive these systems.

ACKNOWLEDGEMENTS

Thanks are due to H.M.J. Udo, L.O. Fresco, M.N.M. Ibrahim, D. Zoebl, P. Hoekstra, G. Montsma and B.H. Slicher van Bath and many others for their comments, and to B.A. Williams for the editing.

REFERENCES

Aalfs, H.G., 1934. De rundveeteelt op het eiland Bali. Proefschrift. drukkerij H.J. Smits, Utrecht. 174 pp.

Achterhuis, H.J., 1990. Van moeder aarde tot ruimteschip: humanisme en milieucrisis. Oratie Wageningen, 29 maart 1990, Wageningen, The Netherlands. 40 pp.

Algra H. (ed.), 1978. Kroniek van een Friese Boer, de aantekeningen (1821-1856) van Doeke Wijgers Hellema te Wirdum. Wever, Franeker. 445 pp.

Amir, P. and Knipscheer, H.C., 1989. Conducting on-farm animal research: procedures and economic analysis. Winrock International Institute for Agricultural Development. 244 pp.

Anon, 1974. Ere-doctor P.B. de Boer: Forketteraers ha hwat tsjineffekt krigen.... Leeuwarder Courant, 9 maart 1974.

Backer, C.A., 1934. Onkruidflora der Javasche Suikerrietgronden, Handboek ten dienste van de suikerriet-cultuur en de rietsuiker-fabricage op Java, zevende deel, aflevering 4, p. 821-807.

Bakker, J., 1973. De produktiviteitsontwikkeling van een intensief melkveebedrijf. Een analyse van de technische en economische ontwikkelingen in de afgelopen 25 jaar van het bedrijf van de familie de Boer te Stiens. 26 pp.

Bech H., Denecke D. and Jankuhn H. (eds.), 1980. Untersuchungen zur eisenzeitlichen und frühmittelalterlichen Flur in Mitteleuropa und ihrer Nutzung: Bericht über die Kolloquien der Kommission für die Altertumskunde Mittel- und Nordeuropas in den Jahren 1975 und 1976. 2 Tl. Van Der Hoek & Ruprecht, Goettingen, Germany.

Beets W.C., 1990. Raising and Sustaining Productivity of Smallholder Farming Systems in the Tropics. AgBé Publishing, P.O. Box 9125, 1800 GC Alkmaar, Netherlands

Bieleman, J., 1987. Boeren op het Drentse zand 1600-1910. Een nieuwe visie op de 'oude' landbouw. Ph.D. Thesis, Wageningen / Utrecht, The Netherlands. 834 pp.

Bieleman, J., 1992. Geschiedenis van de landbouw in Nederland, 1500-1950: veranderingen en verscheidenheid. Meppel: Boom. 423 pp.

Biewinga, E.E., Aarts, H.F.M., and Donker, R.A., 1992. Melkveehouderij bij stringente normen, bedrijfs- en onderzoeksplan van het Proefbedrijf voor Melkveehouderij en Milieu. September 1992, De Marke, Hengelo, Rapport no.1. 284 pp.

Blaisdell, J.D., 1994. The curse of the Pharaohs: anthrax in Ancient Egypt. ARGOS, bulletin van het veterinair historisch genootschap, 10: 311-314.

Bodson, L., 1994. Ancient views on pests and parasites of livestock. ARGOS, bulletin van het veterinair historisch genootschap, 10: 303-310.

Bromley, D., 1992. Making the commons work: theory, practice and policy. ICS-Press, San Francisco, USA, 339 pp.

Buckminster Fuller, R., 1975. Ruimteschip Aarde: een blauwdruk om te overleven = operating manual for spaceship Earth. Den Haag: Bakker, 125 pp.

Campbell, B.M.S. and Overton, M. (eds.), 1991. Land, Labour and Livestock: historical studies in European agricultural productivity. Manchester University Press, Manchester and New York. 500 pp.

Carlier, H., 1987. Understanding traditional agriculture. Bibliography for development workers. ILEIA, Leusden, The Netherlands. 114 pp.

Chambers, R., Pacey, A., and Thrupp, L.A., 1989. Farmers first: farmer innovation and agricultural research. London: Intermediate Technology Publications. 218 pp.

Cochrane, W.W., 1979. The development of American agriculture, a historical analysis. University of Minnesota Press, Minneapolis, 464 pp.

Conway, G.R. and E.B. Barbier, 1990. After the Green Revolution, Sustainable agriculture for Development, Earthscan Publications Ltd. London, 205 pp.

Cooke, 1974. Spaceship travel. In: Cox, G.W., and Atkins, M.D., (eds.), 1974. Agricultural ecology, an analysis of world food production systems. W.H. Freeman and Company, San Francisco.

Crosby, A.W., 1986. Ecological Imperialism, the Biological Expansion of Europe (900-1900). Cambridge University Press. Cambridge, U.K. 368 pp.

Crotty, R., 1980. Cattle, Economics and Development. Commonwealth Agricultural Bureaux, Farnham Royal, Slough SL2 3BN, UK. 253 pp.

Daly, H.E., and Cobb, J.B., 1990. For the common good. Redirecting the Economy towards Community, the Environment and a Sustainable Future. Green Print, 482 pp.

Dawkins, R., 1982. The extended phenotype, the long reach of the gene, Oxford University Press, Oxford / New York, 307 pp.

Dawkins, R., 1992. The blind Watchmaker. Penguin, London, UK. 340 pp.

De Boer, A.J., Singh, C.B., Singh, L., Dixit, P.K. and Patil, B.R., 1994. Rapid Rural Appraisals to Assist Dairy Production in India. - Their use in focusing research. Technical Publication Number 2. ICAR/BIOCON/WINROCK INTERNATIONAL. Indian Council of Agricultural Research, Animal Sciences, Krishi Bhavan, New Delhi, India; Dept. Animal Production Systems, Agric. Univ. Wageningen, The Netherlands.

De Vries, J., 1994. De bestrijding van de runderpest in Friesland gedurende de 18e eeuw. ARGOS, bulletin van het veterinair historisch genootschap, 10: 315-323.

De Wit, C.T., 1968. Plant Production. Reprinted from: Miscellaneous papers Landbouw Hogeschool Wageningen, 1968, No.3, pp. 25-50.

De Wit, C.T., 1969. Een bodemvruchtbaarheidstheorie uit de eerste helft van de 19e eeuw. Landbouwkundig Tijdschrift, 81: p.245-251.

De Wit, C.T., 1992. Resource Use efficiency in Agriculture. Agricultural Systems, 40: 125-151.

De Wit, J., Dhaka, J.P. and Subba Rao, A., 1993. Relevance of breeding and management for more or better straw in different farming systems. pp.404-414. In: Kiran Singh and Schiere, J.B. (eds.), 1993. Feeding of ruminants on fibrous crop residues. Aspects of treatment, feeding, nutrient evaluation, research and extension. Proc. of a workshop, 4-8 february 1991, NDRI-Karnal. ICAR, Krishi Bhavan, New Delhi, India; Dept. Trop. Anim. Prod., Agricultural University, Wageningen, The Netherlands. 486 pp.

DGIS, 1987. Assisting Livestock Development, Experience of development Cooperation with reference to livestock in the period 1978 - 1984. Evaluation Report Netherlands Development Cooperation. 80 pp.

Dijksterhuis, E.J., 1975. De mechanisering van het wereldbeeld. 2e druk. Amsterdam, 590 pp.

Durning, A.B., and Brough, H.B., 1991. Taking Stock: Animal Farming and the Environment. World Watch Paper 103.

Elogoet, F. and Van Gils, L., 1989. Agriculture en Hollande: L'intelligence efficace. Plabennec: Tud HaBro (fr.), 214 pp.

Farrington, J., and Martin, A.M., 1988. Farmer participatory research: a review of concepts and recent fieldwork. Agric. Admin. & Extension, 29: 247-264.

Fraser, A., 1949. Sheep husbandry, Crosby Lockwood & Son Ltd, London, 297 pp.

Fresco, L.O., 1986. Cassava in shifting cultivation, A systems approach to agricultural technology development in Africa. Ph.D. Thesis, Agricultural University Wageningen, Royal Tropical Institute - The Netherlands. 240 pp.

Fresco, L.O. and Westphal, E., 1988. A hierarchical classification of farm systems. Expl. Agric., 24: 399-419.

Geertz C., 1963. Agricultural involution, the process of ecological change in Indonesia. Published for the Association of Asian Studies by University of California Press, Berkeley/ Los Angeles/London. 176 pp.

Gibson, T., 1987. A ley farming system using dairy cattle in the infertile uplands, Northeast Thailand. World Animal Review, 61: 36-43.

Gleick, J., 1987. Chaos, making a new Science. An Abacus Book, London, U.K. 352 pp.

Gryseels, G., 1988. Role of livestock on mixed smallholder farm in the Ethiopian highlands. A case study from the Baso and Worena werada near Debre Berhan. Ph.D. Thesis, Agricultural University Wageningen, Dept. of Trop. Animal Production, Wageningen, The Netherlands. 249 pp.

Hart, R.D., 1982. An Ecological Systems Conceptual Framework for Agricultural Research and Development. p.44-58. In: Shaner W.W., Philipps P.F. and Schmehl W.R., (eds.), 1982. Readings in Farming Systems Research and Development. Westview special studies in agriculture / aquaculture science and policy. Westview Press, Boulder, Colorado. 175 pp.

Hayami, Y. and Ruttan, V.W., 1985. Agricultural Development, An International Perspective. The John Hopkins University Press, Baltimore and London. 506 pp.

Hildebrand P.E. and Waugh, R.K. 1986. Farming System Research and Development. In: Hildebrand P.E. (ed), Perspectives on Farming Systems Research and Extension, Lynne Rienner Publishers, Inc., Boulder, Colorado, USA.

Hoekstra P., 1948. Paardenteelt op het eiland Soemba. Proefschrift. Universiteit van Indonesië, Batavia. 188 pp.

Hopkins T.K. and Wallerstein I., 1982. World-systems analysis: theory and methodology. Exploration in the world economy, Vol. I. Sage, Beverly Hills. 200 pp.

Huitema, H., 1982. Animal Husbandry in the Tropics, its Economic Importance and Potentialities, Studies in a Few regions of Indonesia. Communication 73, Dept. of Agricultural Research, Royal Tropical Institute Amsterdam, The Netherlands, 313 pp.

Joshi, A.L., Doyle, P.T. and Oosting, S.J., (eds.), 1994. Variation in the quantity and quality of fibrous crop residues. Proceedings of a national seminar held at BAIF Development Research Foundation, Pune (Maharashtra-India), February 8-9, 1994. BAIF, Pune; ICAR, Krishi Bhawan, New Delhi, India. 174 pp.

Kang, B.T., Reynolds, L., and Atta-Krah, A.N., 1990. Alley farming. Advances in Agronomy, 43: 316-359.

Keuning, J.A., 1987. Stikstofproefbedrijven als schakel tussen praktijk, voorlichting en onderzoek. p. 55-58. Jaarverslag van Proefstation voor de Rundveehouderij, Schapenhouderij en Paardenhouderij, 1987.

Kiran Singh and Schiere, J.B. (eds.), 1993. Feeding of ruminants on fibrous crop residues. Aspects of treatment, feeding, nutrient evaluation, research and extension. Proceedings of a workshop held from 4-8 February 1991 at NDRI-Karnal. ICAR (Animal Sciences), New Delhi, India. 486 pp.

Kumar, M.N.A., Singh, M. Bhaskar, B.V. and Vijayalakshmi, S., 1993. An economic evaluation of urea treatment technology of straw on farms - a review. p.297-305. In: Kiran Singh and Schiere, J.B. (eds.), 1993. Feeding of ruminants on fibrous crop residues. Aspects of treatment, feeding, nutrient evaluation, research and extension. Proceedings of a workshop, 4-8 february 1991, NDRI-Karnal. ICAR, New Delhi, India. 486 pp.

Lewin, R., 1993, Complexity. Life on the edge of chaos. Phoenix, London, U.K. 208 pp.

Longman, 1985. Longman Concise English Dictionary. Longman, Avon, U.K. 1651 pp.

Lord Ernle, 1961. English Farming, Past and Present, 6th edition, Heinemann, London/Melbourne/Toronto, Frank Cass& Co, London. 559 pp.

Lovelock, J.E., 1979. Gaia, a new look at life on Earth. Oxford University Press, Oxford. 154 pp.

Mahendra Singh, Amrit Kumar, M.N., Rai, S.N. and Pradan, P.K., 1993. Urea-ammonia treatment of straw under village conditions: reasons for success and failure. p.289-296. In: Kiran Singh and Schiere, J.B. (eds.), 1993. Feeding of ruminants on fibrous crop residues. Aspects of treatment, feeding, nutrient evaluation, research and extension. Proceedings of a workshop, 4-8 february 1991, NDRI-Karnal. ICAR, New Delhi, India. 486 pp.

Marwar J., 1989. Farming Systems Research in Indonesia: its evolution and future outlook, p.6-36. In: Sukman S., Amir P. and Mulyadi D.M. (eds.), 1989. Developments in Procedures for Farming Systems Research: Proceedings of an International Workshop, held in March 13-17, 1989, Puncak, Bogor Indonesia. Agency for Agricultural Research and Development, Jakarta, Indonesia.

Merkens, J., 1927. Bijdrage tot de kennis van den karbouw en de karbouwteelt in Nederlands Oost-Indië (Contributions to the knowledge of the buffalo and its husbandry in The Netherlands East-Indies). Ph.D. Thesis. University of Utrecht, The Netherlands.

Merril Sands, D., 1986. Farming systems research: clarification of terms and concepts. Expl.Agric., 22: 87-104.

Nordblom, T., and Halimek, H., 1982. Lentil crop residues make a difference. Lentil Expl. News Service, Newsletter No. 9, p.8-10. Communication Department, ICARDA, P.O. Box 5466, Aleppo, Syria.

Norman, D.W. and Gilbert, E., 1982. A General Overview of Farming Systems Research. chapter 2. In: Shaner W.W., Philipps P.F. and Schmehl W.R., (eds.), 1982. Readings in Farming Systems Research and Development. Westview special studies in agriculture / aquaculture science and policy. Westview Press, Boulder, Colorado. 175 pp.

Nou, J., 1967. Studies in the Development of Agricultural Economics in Europe. Almquist & Wiksells Boktryckeri AB, Uppsala. 611 pp.

NRC, 1981. Food, fuel and fertilizer from organic wastes. Report of an ad-hoc panel of the Advisory Committee on Technology Innovation, Board on Science and Technology for International Development, Commission on International Relations, National Research Council, Washington D.C.

Odum, H.T., 1971. Environment, power and society. Wiley Interscience, New York. 331 pp.

Odum, H.T., 1983. Systems ecology, an introduction. Wiley Interscience, New York. 644 pp.

Olson, M., 1971. The logic of collective action, Public Goods and the Theory of Groups, Harvard University Press, Cambridge/Massachusetts/London/England. 186 pp.

Osinga, A., 1992. Dairy farming systems research in The Netherlands. pp.105-111. In: Gibon, A. and Matheron, G. (eds.), 1992. Agriculture. Agrimed research programme. Global appraisal of livestock farming systems and study of their organizational levels: concepts, methodology and results. Proc. of a symposium organized by the INRA-SAD and the CIRAD-IEMVT, 7 July 1990, Toulouse, France. 510 pp.

Ponting, C., 1991. A Green History of the World, Penguin Books, London. 432 pp.

Posner J.L. and Gilbert E., 1991. Sustainable agriculture and farming systems research teams in semi-arid West Africa: a fatal attraction? Journal for Farming System Research / Extension, 2, (1): 71-86.

Prigogine I., and Stengers I., 1985. Order out of Chaos, man's new dialogue with nature. Flamingo, London, U.K. 349 pp.

Randhawa, M.S., 1980. A history of agriculture in India. Volume I, Beginning to 12th century. Indian Council of Agricultural Research, New Delhi, India. 541 pp.

Randhawa M.S., 1982. A history of Agriculture in India. Volume II, Eight to Eighteenth Century. Indian Council of Agricultural Research, New Delhi, India. 358 pp.

Randhawa M.S., 1983. A history of Agriculture in India. Volume III, 1757-1947. Indian Council of Agricultural Research, New Delhi, India. 422 pp.

Randhawa M.S., 1986. A history of Agriculture in India. Volume IV, 1947-1981. Indian Council of Agricultural Research, New Delhi, India. 716 pp.

Reed, J.D., Capper, B.S., and Neate, P.J.H. (eds.), 1988. Plant breeding and the nutritive value of crop residues. Proc. of a workshop held at ILCA, Addis Ababa, Ethiopia, 7-10 December, 1987. ILCA, Addis Ababa, Ethiopia. 334 pp.

Reijntjes, C., Haverkort, B. and Waters-Bayer, A., 1992. Farming for the future. An introduction to Low-External-Input and Sustainable Agriculture. McMillan, ILEIA. 250 pp.

Rifkin, J., 1989. Entropy into the greenhouse world. Bantam Books, New York, revised edition. 354 pp.

Rifkin, J., 1992. Beyond beef: the rise and fall of the cattle culture. A Dutton book, Penguin, New York. 353 pp.

Rose, H.J., 1954. A Handbook of Latin Literature, from the earliest times to the death of Saint Augustine, 3rd edition, Methuan & Co Ltd, London.

Ruthenberg, H., 1980. Farming Systems in the Tropics. third edition. Clarendon Press, Oxford, U.K., 424 pp.

Scheltema, A.M.P.A. 1927/28. Nieuwe opvattingen over landbouw geografie (New concepts on Agricultural Geography), Landbouw 1927/28: 63-105.

Schiere, J.B., 1974. Men eat pigs or pigs eat men. (Orang makan babi atau babi makan orang) Some problems on the nutrition of pigs under backyard farming conditions in the tropics. Special reference to Central Java and South East Asia. MSc.-thesis, Dept. Animal Nutrition, Agricultural University Wageningen, The Netherlands. 64 pp.

Schiere, J.B., 1991. Livestock systems research in the tropics. A review of dutch experiences. Paper presented at the global workshop on Animal Production Systems, 16-21 September 1991, San José, Costa Rica, IDRC-Ottawa Canada and WINROCK, Morrilton, Ark., USA.

Schiere, J.B., 1993. Research and extension in livestock development. p.135-145. In: Gill, M., Owen, E., Pollot, G.E. and Lawrence, T.L.J. (eds.), 1993. Animal Production in Developing Countries. Proceedings of symposium held at Wye, September 1991. Occasional Publication No. 16 - The British Society of Animal Production (BSAP) 1993, 243 pp.

Shaner, W.W., Philipps, P.F. and Schmehl, W.R., 1982. Farming Systems Research and Development. Guidelines for developing countries. Westview special studies in agriculture / aquaculture science and policy. Westview Press, Boulder, Colorado. 414 pp.

Simmonds, N.W., 1986. A short review of farming systems research in the tropics. Exp.Agric., 22: 1-13.

Slicher van Bath, B.H., 1958. Een Fries landbouw bedrijf in de tweede helft van de zestiende eeuw. Agronomisch hist. bijdr., 4: 67-130.

Slicher van Bath, B.H., 1963. The agrarian history of Western Europe, A.D. 500-1850. Translated by O. Ordish. Arnold, London. 364 pp.

Slicher van Bath, B.H. 1979, Spaans Amerika omstreeks 1600. Aula 50, Het Spectrum, Utrecht / Antwerpen. 272 pp.

Spedding, C.R.W., 1988. An Introduction to Agriculture Systems, 2nd Edition, Elsevier Applied Science, London. 189 pp.

Stroosnijder, L., and Van Rheenen, T., 1993. Making farming systems analysis a more objective and quantitative research tool. p.341-353. In: Penning De Vries, F.W.T. *et al.*, (eds.), 1993. Systems Appraoches for Agricultural Development. Kluwer Academic publishers, The Netherlands. 542 pp.

Struif Bontkes, T., 1992. On Dinka, cattle and grain, an interdisciplinary simulation study of a rural area in Southern Sudan. Wageningen Agricultural University Papers 91-3.

Sumberg, J.E., and Atta-Krah, A.N., 1988. The potential of alley farming in humid West Africa - a re-evaluation. Agroforestry Systems, 6: 163-168.

Sundstøl, F., and Owen, E., (eds.), 1984. Straw and other fibrous by-products as feed. Developments in Animal and Veterinary Sciences, 14. Elsevier, 604 pp.

Swanson, B.E. and Claar, J.B., 1984. The history and development of Agricultural extension, Chap. 1 In: Swanson B.E. (ed.), Agricultural extension, a reference manual, 2nd edition, FAO, Rome, 262 pp.

Thorner, D., Kerblay B. and Smith R.E.F., 1966. A.V. Chayanov on the Theory of Peasant Economy. Published for the Americam Economic Association, by R.D. Irwin Inc., Homewood, Illinois. 317 pp.

Tiesema, K., 1980. De benutting van het grasland op het bedrijf van B.P. de Boer te Stiens. pp. 408-414. Stikstof: mededelingen van het Landbouwkundig Bureau der Nederlandse Stikstofmeststoffen Industrie, 1980, deel: 95/96.

Toulmin, C., 1984. The allocation of resources to livestock research in Africa. LPU working paper no.3, International Livestock Centre for Africa (ILCA), Addis Abbeba, Ethiopia. 66 pp.

Tripp R., Anandajayasekaram P., Byerlee D. and Harrington L.W., 1991. FSR: Achievements, Deficiencies and Challenges for the 1990's. J. Asian Farm. Syst. Assoc., 1: 259-271.

Trow-Smith, R., 1957. A History of British Livestock Husbandry to 1700, Routledge and Kegan Paul, London, U.K.. 286 pp.

Trow-Smith, R., 1959. A History of British Livestock Husbandry 1700-1900. Routledge and Kegan Paul, London, U.K.. 351 pp.

Van Der Krogt, P.C.J., 1992. Franciscus Verburch en zijn 'neef' Cornelis, twee Westlandse Pastoors en hun Familie. p.72-92. In: Historische Jaarboek Westland 1992.

Van Der Ploeg, J.D., 1994. Styles of farming: an introductory note on concepts and methodology. p.7-30. In: J.D. Van Der Ploeg and A. Long (eds.), 1994. Born from within: practice and perspectives of endogenous rural development. Van Gorum, Assen, The Netherlands, 298 pp.

Veeneklaas, F.R., Cissé, S., Gosseye, P.A., Van Duivenbooden, N., and Van Keulen, H., 1991. Competing for limited resources: the case of the fifth region of Mali, Development scenarios report 4. CABO-DLO, Wageningen, The Netherlands, and ESPR, Mopti, Mali, March 1991.

Veldink J.G., 1970. W.C.H. Staring 1808-1877. Geoloog en landbouwkundige. Ph.D. Thesis. Agricultural University Wageningen. Pudoc, Wageningen, The Netherlands. 206 pp.

Vroon, P.A., 1989. Tranen van de krokodil over de te snelle evolutie van onze hersenen. Ambo, Baarn, The Netherlands.

WCED, World Commission on Environment and Development (WCED), 1987. Our common future. Oxford University Press, Oxford, U.K. 400 pp.

White K.D., 1970. Roman Farming, Thames & Hudson, London. 536 pp.

Williamson G. and Payne W.J.A., 1965. An introduction to animal husbandry in the tropics, 2nd edition, Tropical Agriculture Series, Longman. 447 pp.

Chapter 2.2

LIVESTOCK AND FARMING SYSTEMS RESEARCH
II: DEVELOPMENT AND CLASSIFICATIONS

J.B. Schiere and J. De Wit

SUMMARY

Farming systems change on scales of space and time. A change on the latter is called development, and can be both the cause and result of innovations in technology and management. The development process is often implicitly equated with progress, but that notion is challenged here as a background for the discussion of the use of crop residues as animal feed. An understanding about the role of straw as livestock feed in a large variety of conditions and technologies requires a classification of farming systems, preferably based on criteria that determine system behaviour. A two-dimensional matrix is therefore proposed in which the vertical axis represents the relative access to the production factors land, labour and capital. An underlying distinction between open and closed systems, i.e. high *versus* low input systems is particularly important for the discussion of system behaviour in relation to straw feeding. The horizontal axis reflects the degree of interactions between crops and livestock: from almost pure livestock on the left, via mixed crop-livestock systems in the centre towards predominantly cropping on the right. The matrix serves as a framework for the discussion on the usefulness of straw feeding systems in subsequent chapters. Special attention is given to the characteristics of mixed farming in on-farm and between-farm situations, as they are found on a scale from diversification to integration. Differences between systems require strategies for development, criteria for system evaluation and straw feeding technology that are adjusted to the conditions of the system concerned.

INTRODUCTION

The discussion about the use of straw as cattle feed on a more than site-specific level requires a classification of farming systems. Such classifications have been given by many workers in tropical and temperate areas, e.g., Duckham and Masefield (1970), Ruthenberg (1980), Jahnke (1982), Nestel (1984), De Boer (1985), Grijseels (1988), Simpson (1988)and 't Mannetje (1989).

The most relevant classifications are based on factors that explain and/or drive system behaviour, providing an opportunity to interpolate and extrapolate about system behaviour. *Ad hoc* experimentation can then be avoided and hypotheses can be formulated for subsequent testing and theory development. For example, Herlemann (1954), Crotty (1980) and Pingali *et al.* (1987) explain system behaviour based on access to resources, i.e. the relative cost of production factors, combined with demand for produce, hereafter called resource / demand patterns. In that sense, they provide a follow-up of the work by Von Thünen on locations of agriculture some one hundred and fifty years ago (Nou, 1967; Ch. 2.1).

This chapter - the second in a series of three - first briefly discusses the development of agricultural systems. It explains the role of technological innovation, the implicit notion of progress in development, and the criteria for system success and classification, as well as the difficulty of classification in general. After that a matrix of farming systems is proposed based on resource / demand patterns, and the nature of interaction between crops and livestock. Apparently unrelated systems such as modern high input urban dairy, and a traditional low input draught animal system are thus placed in a two-dimensional space. The matrix is designed to understand and to interpolate the role of straw as feed in livestock systems, i.e. about the usefulness of straw feeding methods in different systems. The paper is based on the first chapter that reviews backgrounds and terminology of FSR. The next chapter describes the role of straw for feed and other purposes in mixed farming systems.

AGRICULTURAL DEVELOPMENT

Development of systems can be assumed to be determined, mostly if not entirely, by what we call **resource / demand** patterns. These patterns express access to technology and management, as well as different value systems. (Odum, 1971; Harris, 1974, 1985 and 1988; Crotty, 1980; Hayami and Ruttan, 1985; Pearson, 1992; Smith *et al.*, 1993; Van Der Ploeg and Long, 1994). Development is diverse in appearance and reasons, being both the cause and effect of changing resource / demand patterns (Slicher Van Bath, 1963; Boserup, 1965; Grigg, 1974, 1982; Noy-Meir and Seligman, 1979; Hayami and Ruttan, 1985). The main terminology will be defined first to clarify further discussions.

Terminology

The term **development** is defined by Longman (1985) as: the act, process or result of developing. The verb 'to develop' has several meanings according to the same dictionary. We prefer the definition 'to go through a process of natural growth, differentiation, or evolution by successive changes'. It acquires the meaning of morphogenesis, and it is here used in a neutral sense. It does not imply a one-way direction on a ladder to an imaginary higher goal. **Resources** are considered to consist of the classical production factors: land, labour and capital. **Land** is an aggregate of land quantity and quality, including aspects such as soil fertility, water and climate. Access to land is directly -though not linearly- related to access to plant biomass. **Labour** is an aggregate of individual skill and number of persons. **Capital** refers in this and the following chapters to inputs such as fertilizer and commercially compounded feeds. Other forms of capital (e.g. cattle) are either derived from access to land, labour and inputs, or incorporated into the value of land (e.g. wells, fences, irrigation infrastructure) or labour (e.g. education). The effect of the **market** is captured under the term **effective demand**, which can be defined as 'the product of number of people and per capita consumption',where per capita consumption refers to the use of products such as food, fibre, wine, as well as, ultimately products such as building materials, and fuels. The effect of the market is directly related to access to capital inputs: without commercialization there is no possibility to purchase inputs! **Innovation** stands for changes in management, technology and institutions (Hayami and Ruttan, 1985). The use of terms such as **'effective demand'** and **'access to'** rather than **'availability of'** resources indicate that some subsystems - whether societies or individuals - consume or control a larger share of the resources than others. This is expressed in the terms 'pressure of people on people' *versus* 'pressure of people on resources' (White, 1976). This principle will be referred to later on where it is stressed that the need for development can come from within, as well as from outside the system: a crucial effect of boundary conditions. Access to food and resources can improve due to better distribution, even for a small or no increase of total production (Amartya Sen, 1981).

System adjustments

Adjustments to decreasing resource / demand ratios have been reviewed by Grigg (1974; 1982) and Palthe (1989). They are here categorized as:
- expanded land use,
- change of consumptive habits,
- use of innovations in technology and management.
Changing value ratios between land, labour, capital and demand are likely to be the driving forces behind system behaviour, and they provide the basis of the classification in this paper.

Expanded land use occurs due to emigration, in the cultivation of more land, or of the same land more frequently. Inadequate access to land relative to effective demand, has driven many migratory movements throughout the history of the world, including the migrant labour and ecorefugees of today (Crosby, 1986; Ponting 1991; Kaplan, 1994). Mining of fossil deposits and natural fertility is a modern, but disguised form of expansion: into the earth and into the future (Meadows *et al.*, 1972; WCED, 1987; Van Der Pol 1992).

Change of consumptive habits is an adaptation of human as well as animal subsystems. Food quality (protein) is generally traded for quantity (carbohydrates) in conditions of scarcity. The German expression **'Vergetreidung'** (='graining') indicates a move from food of animal origin to grain in medieval Europe (Roscher 1888, quoted by Bieleman 1990). In many systems, pulses are replaced with grains and grains for tubers (Lagemann, 1977;

Fresco, 1986; Palthe, 1989). Shortage of food - energy for humans in its most basic form - also drives the search for exotic resources: marine foodstocks, single cell protein from oil, or domestication of hitherto 'unexploited' species of animals or plants. It also leads to the need for the development of new farming systems, called NFSD in Ch. 2.1. The true nature of this process surfaces in titles of documents that focus on *'exploitation of new'* resources (FAO, 1977; NRC, 1981; NRC, 1991). The research on tree leaves and so-called 'unconventional feeds', is a further symptom of an attempt at expansion. A shortage of wood or charcoal causes people to use dung cakes for fuel, in today's Asia as well as in the Britain of the 18th century. Lord Ernle (1961) said:

both in Buckinghamshire and in Northamptonshire, the cow-dung was collected from the fields, mixed with short straw, kneaded into lumps, daubed on the walls of buildings, and, when dry, used as fuel.

The need to focus attention on the use of straws as animal feed - the topic of this thesis - is a good example of livestock adjusting its consumptive habits: away from bush and range grazing towards use of crop residues, whether or not 'guided' by the farmer.

The *use of innovations in management and technology*, is here summarized under the term technology adoption. Both management and technology are explicitly mentioned since they can be seen to represent use of information and energy for system control, an issue further elaborated in Ch. 6. The importance of management is often forgotten in modern technology-driven transfer of technology (Röling, 1989). It was, however, well-known by older authors. Lord Ernle (1961) quotes one of the early British writers on farming (Googe):

[...] farmers can not thrive by manure [and machinery] alone. On the contrary [...] 'the best doung for ground is the Maister's foot, and the best provender for the house the Maister's eye.'

Inventions can be discovered by chance e.g. penicillin, or after a deliberate search. The development of high yielding grain varieties or straw treatment methods (Sundstøl and Owen, 1984; Kiran Singh and Schiere, 1993) illustrates the active search for what are called, induced innovations by Hayami and Ruttan (1985). The recent work on development of grain varieties with more and/or better straws (Reed *et al.*, 1988; Joshi *et al.*, 1994) fits in the same mould. That adoption of technology is therefore not necessarily a sign of progress, is illustrated in the saying:

'necessity is the mother of invention'.

In fact, the search for and application of innovations can be both the cause and result of shortages (Wilkinson, 1973; Hayami and Ruttan, 1985; Crosby, 1986; Ponting, 1991). Shortage and need are relative concepts, as they can change with access to resources (Achterhuis, 1988).

The change from gathering and hunting, to cultivation and animal husbandry appears generally to require more work at the same or a lower level of nutrition (Cox and Atkins, 1974; Ponting, 1991). Technology is put further into perspective by considering the often externalized negative technical and social side-effects: pollution, resource exhaustion and social disparity or unrest (Meadows *et al.*, 1972; WCED 1987; Rifkin, 1989; Conway and Barbier, 1990; Ellul, 1990). Technology can also effectively drain the system even more, reducing long term sustainability. Crops, animals and management that survive under low input conditions often extract the last resources. 'Uitmergelen' is a Dutch word for the excessive application of 'marl' (Slicher Van Bath, 1963). Whereas 'marling' appeared to act as a fertilizer, in reality it led to further depletion of resources, i.e. by releasing bound soil phosphate and/or by speeding up the breakdown of soil organic matter on acid soils. The results of accelerated mining by marl technology is expressed by a quote from Googe (Lord Ernle, 1961) who cautions against the persistent use of chalk, because, in the end:

'it brings the grounde to be starke nought, whereby the common people have a speache, that grounde enriched with chalke makes a riche father and a beggerly sonne.'

In more modern English:

'Lime and lime without manure,
makes both land and farmer poor'
(G. Montsma, pers. comm. 1993).

CLASSIFICATION OF SYSTEMS

Universal classifications are either too clumsy or too general, and the quest for precision increases complexity (Traub and Wozniakowski, 1994). For practical purposes, it is therefore necessary to suit the classification criteria to the objective and conditions of the study. A useful classification employs criteria that determine system behaviour. Such criteria can consist of what ecologists call **indicator processes** or **indicator species**. Others term them **proxy variables**, since they represent a system, rather than being the system itself (Stocking and Abel, 1981). The classification by Ruthenberg (1980) uses indicator processes (i.e. shifting cultivation; fallow; arable irrigation; grazing) as well as indicator species (i.e. dairy and ley; perennial crops). Besides these, there is an infinite range of processes, species or criteria based on agro-ecological, socio-economical and other variables or indicators. This thesis draws mainly upon the approaches by Herlemann (1954), Crotty (1980), and Montsma (1984), implicitly reflecting a classification into open and closed systems.

Open and closed systems

A sharp distinction between open and closed systems cannot be drawn, but their position at extremes of a scale will prove to be useful[1]. The terms respectively reflect high external input agriculture (**HEIA**) and low external input agriculture (**LEIA**). Each type of system has a different behaviour, elaborated for animal nutrition in Ch. 5.1 and 5.2, and in a thermodynamic context in Ch. 6. Simply speaking, and considering only the extremes on the scale, open systems can import resources to satisfy their demands, whereas in closed systems the demand has to be adjusted to the resources. An awareness of closed systems may be lost in modern HEIA systems, but it is well known in LEIA systems, e.g. expressed in folkwisdom (Box 1), and it transpires in recent sociological literature about development in farming systems (Van Der Ploeg and Long, 1994).

If two subsystems operate within one closed larger system, the more powerful subsystem can be considered to be relatively more open than the weaker one. The stronger can

---

[1] In thermodynamic terms it is probably better to speak of systems distant from, and near equilibrium (Ch. 6).

import resources from the weaker, either directly or via the market. The stronger system can also force (externalize) its surpluses into the weaker systems. In many mixed crop-livestock systems, animals as a weak subsystem, have to adjust to crops more than vice versa, a central issue in this thesis and expressed by a farmer in West Bengal (India):

> 'why should I waste mustard oil seed cake on animal feed if I can use it to fertilize my valuable crop'.

In systems where more income is derived from animal produce than from crops, it is obviously the crop system that needs to adjust to the animals. For example, in the Netherlands, farmers from one farming system would learn at school that 'animals serve the crops', whereas farmers from other systems learned that the crops serve the animals (pers.obs.).

The relation between thinking in terms of closed *versus* open systems, is reflected in the tension between holistic and reductionist approaches. The former treats system boundaries and externalities differently than the latter. Output of an individual subsystem that exceeds the resource endowment of a closed system is what we call a damning objective in Ch. 5.1. Damning objectives are realized at the expense of resources from another system, or else, they result in no production, or at least greatly reduced output.

Box 1:   AWARENESS ABOUT THE PRINCIPLE OF CLOSED SYSTEMS, EXPRESSED IN FOLK WISDOM AND CULTURE.

The awareness of closed system conditions is expressed in folk sayings and sociological / anthropological behaviour. A further analysis might refine the interpretation, but here it is worthwhile to note sayings such as:

- *if it can't be done as it should, it should be done as it can*
- *it should be cut from the length or from the breadth*

Cultural behaviour is at least to some extent, determined by an awareness about closed systems. Many social mechanisms that govern exploitation of common resources explicitly restrict the individual's level of consumption in order to sustain the entire community (Wilkinson, 1973; Bromley, 1992). Interesting phenomena in this respect are:

- shared poverty
- the image of the limited goods

Shared poverty in its strictest form implies that society imposes a limit on consumption and wealth accumulation within classes by 'borrowing' excess wealth from emerging wealthy members (Geertz, 1963; Cancian, 1989). The definition of the image of the limited goods implies that broad areas of peasant behaviour are patterned so as to suggest that peasants view their social, economic, and natural universes - their total environment - as one in which all the desired things in life such as land, wealth, health, friendship and love, manliness and honour, respect and status, power and influence, security and safety, *exists in finite quantity and are always in short supply*, as far as the peasant is concerned. Not only do these and all other 'good things' exist in finite and limited quantities, but in addition *there is no way directly within peasant power to increase the available quantities* (Foster, quoted by Cancian 1989).

The relation between two subsystems can be seen in a more conciliatory light, when considered in plantphysiological or ecological terms of sink and source (Warren Wilson, 1972). For example the roots of a tree are a source, and the leaves a sink, for water and minerals, whereas the roots are a sink, and the leaves a source, of carbohydrates. The sink/source relationship is found in mixed crop-livestock systems where crops are the source of straw, but sink for draught, and the animals are a sink for straw and a source for draught. In that sense, two subsystems can be mutually supportive, provided they are adjusted to each other (Patil *et al.*, 1993). In relation to the subject of this thesis: in closed systems the problem of poor quality feed can be overcome by adjusting the animal production level to the resources. In open systems, the feed resources are adjusted to the desired production level (Ch. 5.1. and 5.2).

A TWO WAY MATRIX FOR CLASSIFICATION OF LIVESTOCK SYSTEMS

A classification of livestock systems can be based on the assumption that resource / demand patterns significantly determine system behaviour, i.e. it must be possible to understand system behaviour by using those patterns as an explaining variable. We have chosen two scales to achieve this objective, and in combination they form the matrix of Table 1. The vertical axis consists of four modes of farming (farming systems), differentiated on the basis of relative access to resources. The horizontal axis contains three classes that explain the degree of interaction between crops and livestock.

This classification constitutes a mix of Von Thünen's 'Standörte' (locations) and the Stufen (stages) from German schools in the last century, more recently by workers such as Rostow (see Hayami and Ruttan, 1985). It uses the traditional production factors of land, labour and capital, as determinants of farm system behaviour, as was also implied in the definition of a farm system by Fresco and Westphal (1988) (see Ch. 2.1). Relative access to the resources is approximated with minuses and plusses in the second column of table 1 as suggested by Herlemann (1954). The horizontal axis reflects the relative importance of crop and livestock in the farming systems. This classification is based on Montsma (1984) who used independent, complementary and competing livestock systems, a classification modified by DGIS (1987) as: independent, mixed and competing crop-livestock systems.

No classification is perfect, but elaboration of this one, would not significantly alter the points to be made. The following comments are however required:
- the broken lines between matrix cells indicate that we allow gradual rather than abrupt transition between modes (see also Box 1 in Ch. 2.1),
- the pluses and minuses indicate relative, and not absolute access to resources within that mode. In a system with a large population, but with even more capital, the labour can be still relatively scarce, e.g. in industrial HEIA systems and
- distinct modes can occur in the course of time, or simultaneously in one region or village, reflecting differences in access to physical resources and demand, as well as sociological differences in styles of farming, i.e. value perceptions,
- any suggestion of a fixed sequence in development is to be avoided at this point! Crotty (1980) describes how British livestock systems shifted back and forth between 'modes' over the centuries. Additional examples are discussed for farming systems in Tanzania by Meertens *et al.* (1994). The principle is discussed also by Boserup (1965) and Grigg (1974, 1982), who both speak of regression of systems, implying a notion of good and bad that has been deliberately avoided here.

Table 1. A matrix for a classification of farming systems for the discussion on crop residue feeding

MODE	RELATIVE ACCESS TO RESOURCES			RELATIVE IMPORTANCE OF CROPS AND LIVESTOCK NATURE OF CROP + LIVESTOCK INTERACTIONS		
	land	lab.	cap.	PREDOMINANTLY LIVESTOCK	MIXED	PREDOMINANTLY CROPS
expansion	+	-	-	- herding of cows, pigs poultry on common lands 11, 3 - nomadism, transhumance 26, 35 - wool, mutton, beef ranches in Australia and USA 34 - grazing in Amazon[a] 8,9, - animals on peat soils, highlands and heavy clays: Scotland, Andes, South Holland 34, 50, 51	- draught based on grazing of common lands 3, 11, 33 - dung from grazing on common land (infield-outfield) 2 - Konzentrationswirtschaft 4 - West African agropastoral systems 5, 36 - Maring pigs / shifting cultivation 49	- shifting cultivation 10, 32 - large scale grain production 6, 17
land shortage mode						
LEIA[c]	-	+	-	- landless animal keeping based on cutting of roadside grasses - involution livestock systems 1	- traditional Portuguese mountain agriculture 46 - straw treatment with kitchen ash 7 - Alpine mixed systems 47 - stall feeding based on roadside grass and cropresidues - intensive dung collection during grazing 13 - single yoke draught 14 - thinning / stripping / intercropping of graincrops for fodder 16	- intensive irrigated rice: involution 55 - vegetable growing (in highlands) with no inputs - horticulture

new conser-vation[c]	-	-/+	-/+	- legume based pastures 27, 41 - New Zealand legume based dairy 28, 19, 34 - fodder banks in Nigeria 45 - De Marke (The Netherlands) 43 - dry season fattening 24	- straw treatment with ammonia or urea 23, 29 - specialized, limited input dairy or pig farming 52, 53 - ley systems 30, 31 - adjusted cropping patterns: Flemish system, Norfolk system 3, 11, 12 - animal crop systems in Mediterranean 34, 36 - specialized, limited input, legume based dairy goats 44 - cereal/legume leys Australia 34, 36 - pigs/feedlots with graingrowing based on store/feeder animals 34 - pigs on sugarcane 37 - Amish farming 38, 42, 47 - alley farming with animals 39 - adjusted cropping patterns 12 - mixed systems of heavy soils in The Netherlands - grassplanting + watercatchment 54 - mixed crop and livestock farm in Pennsylvania 53	- ecological farming 56 - mixed cropping 15 - Eastern UK grain systems 33 - Tree, fruits, walnuts and vegetables in California 53 - Florida fresh-market vegetable production 53 - rice production in California, Lundberg Family Farms 53 - mixed tree and food crop in humid tropics 25 - alley farming in humid West Africa 40, 39 - rice cultivation pre-High Yield Varieties 49
HEIA[c]	-	-	+	- specialized dairy on heavy clay or peat in The Netherlands 36, 51	- dairy or other livestock under coconuts or fruittrees[b] 21, 22, 22a, 34 - cut and carry with fertilized napier 28 - dairy with fertilized fodder on arable soil in The Netherlands, Java 34, 36, 48 - urban dairies, industrialized pig and poultry, feedlots 18, 19, 20	- vegetable horticulture - greenhouse farming - industrial plantations - High Yield Varieties in irrigated grain corps 49

The examples in this matrix are referred to literature by numbers. The most important general reviews are here underlined.

1 Campbell and Overton, (1991); 2 Mc Court (1955); 3 Slicher van Bath (1963); 4 Willerding (1980); 5 Wilson (1986); 6 Gever *et al.* (1986); 7 Ramírez *et al.* (1991); 8 Poelhekke (1984); 9 Hecht (1993); 10 George (1990); 11 Lord Ernle (1961); 12 Patil *et al.* (1993); 13 T. Teunissen, pers. comm. (1989); 14 Grijseels (1988); 15 Altieri (1991); 16 Byerlee *et al.* (1989); 17 Grigg (1974); 18 Walshe (1991); 19 Nestel (1984); 20 Gass and Sumberg (1993); 21 Reynolds (1980); 22 Iniguez & Sanchez (1990); 23 Schiere and Ibrahim (1989); 24 Bartholomew *et al.* (1992); 25 Watson (1983); 26 Jahnke (1982); 27 't Mannetje and Jones (1992); 28 Bryant (1986); 29 Westgaard and Sundstøl (1986); 30 Gibson (1987); 31 Martin (1944); 32 Ruthenberg (1980); 33 Crotty (1980); 34 Duckham and Masefield (1970); 35 Simmons (1989); 36 Pearson (1992); 37 Preston and Murgueitio (1992); 38 Fisher (1978); 39 Kang *et al.* (1990); 40 Sumberg and Atta-Krah (1988); 41 FAO (1991); 42 Kraybill (1993); 43 Biewinga *et al.* (1992); 44 Schiffeleers pers. comm. (1993); 45 Waters-Bayer and Bayer (1987); 46 Van Den Dries en Portela (1994); 47 Netting (1993); 48 Bakker *et al.* (1982); 49 Bayliss-Smith (1991); 50 Brouwer *et al.* (1991); 51 Roep and De Bruin (1994); 52 Francis *et al.* (1990); 53 NRC (1989); 54 Conway and Barbier (1990); 55 Geertz (1963); 56 Reijntjes *et al.* (1992)

Notes   a: Barbed wire is capital used to protect the land, not so much as a production factor for improved animal production (Poelhekke, 1984; Rifkin, 1992)
b: Livestock is often used as a form of cheap labour to control undergrowth.
c: All these three belong to the land shortage mode.

The matrix of table 1 has been presented with a number of practical examples in each cell. They are based on case studies drawn from literature, discussions and personal observation, where possible provided with references. The inclusion of examples in one cell does not imply that they are in the middle of a cell, they may in fact, be halfway between two cells, or even somewhat arbitrarily placed.

THE VERTICAL AXIS: MODES OF AGRICULTURE

A brief discussion of each mode is a basis for the next chapters. The increase of effective demand over land, together with access to inputs, will be seen to be a major driving factor behind the change of agriculture.

The *expansion mode* implies that a local land shortage can be solved by expanding the area under exploitation, a form of migration. The use of inputs (capital) is not yet relevant in these systems, because use of other land is easier, the case of traditional nomadism and shifting cultivation. Typical indicator processes in these systems are colonization and the opening up of new land or migration, which eventually leads to deforestation and/or erosion unless more permanent systems of agriculture are developed (Crosby, 1986; Lockeretz, 1989; Ponting, 1991; Rifkin, 1992). These systems include so-called 'Konzentrazions Wirtschaft' (Willerding, 1980) where livestock provide a way to scavenge large areas (outfields) for the concentration of nutrients and energy to smaller areas of infield (McCourt, 1955; Schiere, 1992). Losses are not counted in these systems since land is sufficient. Strictly speaking, and maybe controversially, these systems are based on a high external input approach. They import soil fertility from outfields, either by fallowing or by grazing. Infield / outfield ratios of 1:20 or higher are not uncommon, neither in shifting cultivation nor in animal based systems (Slicher Van Bath, 1963; Ruthenberg, 1980).

The *land shortage mode* combines the three farming systems with low, medium and high use of inputs that will be discussed below. This mode occurs when expansion can no longer meet effective demand of a growing and/or more demanding population. It can occur even where societies - including some in Europe - use mechanisms such as shared poverty or image of the limited goods to control demand (Box 1). Pastoralists have had rules to control grazing and animal pressure, cropping societies have also controlled the use of 'common lands'. The pressure to open up or to develop common lands does not come only from within society. It has quite often come from outside rather than from inside, (Bromley, 1992) and colonialism occurs in several forms, essentially representing a pressure of outsiders on the land of a given community (Crosby, 1986; Ponting, 1991). Even when farmers were content with their way of life, outsiders have forced them to produce more e.g. through imposition of taxes, or their lands were actually taken over by war or legislation. The move for private ownership of common lands as found in Sub-Saharan regions finds its historical equivalent in the enclosures of the UK and the 'Markewet' of the Netherlands (Lord Ernle, 1961; Slicher Van Bath, 1963). The formalization of common ownership itself was a defense against intruding farmers / communities from outside (Slicher Van Bath, 1963).

Effective demand can be restrained involuntarily by Malthusian effects such as disease or social unrest. Possibly, the medieval black death can be seen as a Malthusian response to

scarce resources (Crotty, 1980), the Irish famine of the 19th century is a typical case of the same problem (Ponting, 1991). Decline of societies can also be related to a declining resource base relative to the effective demand. It often starts on the fringe of a system, not well noted in the centre (Kaplan, 1994). Population growth can be restrained on purpose by birth control, polyandry and delayed marriages, or more drastically, by infanticide or leaving behind of elderly and ill people (Wilkinson, 1973; Grigg, 1974; Crotty, 1980; Crosby, 1986). This approach recognizes the limits to growth: i.e. the principle of closed systems. If effective demand is not controlled, it leaves agriculture with two extreme options or their combination, i.e., to:
- proceed with a shortage of capital and a relative abundance of labour (*low external input agriculture, LEIA*),
- proceed with a system based on the use of capital (inputs) where labour is relatively scarce to inputs (*high external input agriculture or HEIA*).
- a mix of LEIA and HEIA, here called new *conservation agriculture*
The fact that one subsystem can expand in a situation of limited resources for the overall system implies that LEIA and HEIA can exist at the same time, and side by side. The LEIA mode refers to a situation where relatively abundant labour is used to increase or to sustain output from the land. Essentially, it implies the application of refined cultivation methods or individual attention for crops and animals. The process is called *involution* for an archetypal case in labour-intensive irrigated paddy systems of Java (Geertz, 1963), where ever more frequent transplanting of rice and/or elaborate irrigation methods can increase land productivity, even at decreasing marginal returns for labour.

Involution as a form of LEIA is reported for European conditions in systems where increased attention to individual plants or animals compensated for the relative shortage of land on poor soils of marginal areas, e.g. in the Netherlands of previous centuries (Bieleman, 1987). Involution is not possible where land quality and associated labour productivity is too low to sustain a population (Posner and Gilbert, 1991). The labour-intensive Flemish system with stall-fed livestock on deep litter systems of centuries ago can be considered a form of involution (Slicher Van Bath, 1963) since animals served to absorb labour. Preparation of dungcakes is found where shortages of firewood or charcoal are overcome by employing more labour per unit of energy. Stripping or thinning of graincrops for fodder is also a form of involution (Byerlee *et al.*, 1989). Other typical indicator processes of involution or LEIA are careful collection of straws, dung and urine, a marked contrast with HEIA where straw is even burned for easy disposal (Hanley and Lindgard, 1987; Kelley, 1992) and where excess dung becomes a liability. LEIA with animals and abundant labour are careful not to waste nutrients in animal excreta: children or adults collect the excreta as soon as they fall. Farmers from the sandy soils of the Veluwe in the Netherlands around the turn of this century are said to have employed special baskets for dung collection which they carried while herding the animals (T.Teunissen, pers. comm. 1988). Netting (1993) even talks of farmers in Alpine systems that carry eroded soil in baskets back up on the hills, a form of involution!

High external input agriculture (HEIA) compensates for land shortage mainly by use of external inputs, even to the point that land becomes available in excess (WRR, 1992). The price of chemicals and fertilizers in these systems are low relative to labour and value of produce, situations typically described by De Wit *et al.* (1987). HEIA represents *an expansion mode in disguise* because the process constitutes an expansion in time: future and non-renewable resources are used in the present. Typical indicators of HEIA are monoculture, specialization and pollution, e.g. industrial pig and poultry keeping or highly specialized dairying, i.e. with high system control.

The term *new conservation agriculture* is based on concepts of workers such as Young and Thaer in the late 18th and early 19th century (Hayami and Ruttan, 1985). They stressed the need for conservation agriculture when they realized that whatever is taken from the soil should be returned. Our addition of 'new' emphasizes the need to reduce losses, rather than to replenish lost nutrients only. In fact, inputs can be needed to prevent losses of resources, essentially by plugging of leaks. Straw treatment with chemicals as applied in the autarkian Norwegian economy of World War II and thereafter, is a case where relatively small amounts of limited inputs such as alkali were used to avoid waste of available resources such as straw (Westgaard and Sundstøl, 1986). Typical indicators of new conservation agriculture should be based on the principle of closed systems: critical use of non-renewable resources, avoidance of externalities, adjustment of effective demand to the resources, recycling of resources and reduction of losses. It is easier to discuss these systems in normative than in practical terms, though much experimentation is now underway and some success of new farm designs is apparent (NRC, 1989; Biewenga *et al.*, 1992).

THE HORIZONTAL AXIS:
CROP-LIVESTOCK INTEGRATION AND STRAW USE

The use of straws for animal feed is mainly relevant in the mixed crop-livestock systems of the central column in Table 1, where mixing of crops and livestock takes place on and between farm systems. Different systems of straw feeding, according to mode of farming are discussed in Ch. 2.3, first we will discuss the principle of integration from a system point of view. Mixed crop-livestock systems are not always possible nor desirable, depending on the demand patterns of the system. Socio-economic or physical factors may exclude either crop or livestock production in given systems. Cropping is not practical or profitable in the left hand column of the matrix due to lack of rain, poor soils or low population density. On the other hand, disease may limit or prevent animal production in areas of the righthand column, e.g. the case of the tsetse fly that causes trypanosomiasis. Mixed enterprises may also be counterproductive, since they appear to require more management capacity and/or capital, i.e. they may operate against the economies of scale.

The intensity of the crop-livestock mixing in the central column ranges on a scale from diversification to integration. Diversified farm systems consist of independent farm subsystems, e.g. poultry, dairy and pig production that hardly exchange resources and waste such as feed, dung or draught, except for the farmers' labour and cash. Diversification is essentially a way to spread labour and risk, if one sector fails, the other serves as a back-up. At the other end of the scale, integrated farm systems consist, in our vision, of interdependent farm subsystems: the animal eats the crop residues and produces dung and draught power for the crops. Such integration aims to avoid a loss of feed biomass and soil nutrients, i.e. to better recycle resources. It can, however, increase risk by its intricacy: a poor harvest of grain in one season can ultimately mean that the 'cow' cannot pull the plow in the next season.

In an abstract sense, livestock in an integrated system can be seen as an additional crop in a multicropping system that aims to reduce waste, for example by avoiding leaks of nutrients. In multicropping, the crops also have to be adjusted to each other, in terms of light interception, disease, pest and rooting patterns (Altieri, 1991). Maximum output of integrated crop-livestock systems also requires that the subsystems are adjusted to each other, shown in the thought experiments of Patil *et al.* (1993), and by the practical observation of De Vries (1947) who wrote:

> '*in Pasuruan (East Java) the animals live entirely from waste products of cropping and the farmer's backyard. This of course, does not benefit the quality of the animals, but there happens to be no space to feed <u>and</u> men <u>and</u> animals.*'

De Vries recognizes here that the quality of the livestock (he probably refers to their output in terms of milk and meat) has to be adjusted to that of the crop subsystem, in order to achieve maximum total system productivity.

Integration and recycling can lead to the reduction of losses, i.e. making the system more productive in one sense. However, integration can also be expansion in disguise, a principle that we propose to call the **Simon effect**. The effect is named after a cook/gardener Simon who was asked to collect the fallen branches and sticks in the Sri Lankan mixed tree garden of the first author. The latter assumed that the sticks and branches would rot and waste anyway, so why not use it for the construction of a rustic cattle shed under the coconut tree, between the coffee bushes and the pepper vines. Simon needed to be reminded frequently, as it appeared later because he used to take this wood home to supply his wife's kitchen with firewood. In other words, where the first author thought he improved system efficiency by recycling otherwise useless material, i.e. by integrating systems, he was in fact taking away resources from use in another, weaker subsystem.

The Simon effect can thus be defined as:

> the use in one system of a seemingly wasted resource, at the expense of its hidden use in another system.

The Simon effect is a typical case of internalization and externalization, the ecological version of robbing Peter to pay Paul! From a holistic system approach, the higher output of the subsystem is achieved at the expense of another, often weaker, subsystem, i.e. the <u>shortages</u> from the stronger system are externalized.

Straw use for feeding in mixed systems knows a number of Simon effects. Treatment may improve straw quality enough so as to make it useful for the strong farmer to feed it to his/her own cows rather than to give it away (Ch. 4.1). Another question is whether 'straw should be used for feed or fertilizer, i.e. does feeding of straw to animals 'rob the nutrients of the soil microorganisms'? (Budelman and Van Der Pol, 1992). Typically, this is a symptom of increased competition for resources, resulting from a shortage of biomass. The role of straw as animal feed in different modes of mixed farming of the central column is the topic of the third chapter, as a preparation for Section 2, 3 and 4.

Between-farms mixing: herders and cultivators

Good social relations - i.e. symbiosis - can exist between cattle keepers and crop producers due to between farm system integration, but animosity can occur, particularly when access to land becomes limited. Competition for land between nomadic tribes and sedentarized farmers increases, in spite of traditionally good cooperation between them (Powell and Waters-Bayer, 1985; Grijseels 1988). In LEIA or expansion agriculture the pastoralists have traditionally used their animals to scavenge crop land, giving dung in return (George, 1990; Wilson, 1986; Pearson, 1992).

Animosity between herders and croppers is known from ancient stories such as of Cain and Abel (Genesis IV). The Great Wall of China was constructed to protect crop-producing Chinese civilizations against invading pastoralists, and the cropping societies in the Gangetic and Indus plains were overrun by mobile cattle-herding Aryans from the North West (Crotty, 1980; Randhawa, 1980). An elaborate discussion of pastoral systems is beyond the scope of this paper, but they are mentioned here as they have occasional use for straws as feed in emergencies. When straw is available and in spite of its low nutritive value, it can become essential for herd survival, either by transporting feed to the animals or by taking the animals to the feed available on crop farms. The exchange of land between potato or flower bulb growers and grass farmers or fodder maize producers in the Netherlands is an example of between farming mixing in HEIA (Anon, 1981).

On-farm mixing, a symbiosis?

On-farm mixing of crops and livestock occurs in systems with limited access to fossil fuel based inputs, where high population densities are combined with cropping and/or where feed can be 'imported', sometimes by grazing on off-farm wastelands. A positive symbiosis between people and animals in those systems seems to take place, e.g. on Java where two thirds of the human population as well as two thirds of the national animal population is concentrated on about one tenth of Indonesia's land surface (Tillman, 1981). India and Bangladesh also have high human and livestock populations in Asia (Barton, 1987) and similar systems are known elsewhere in the world, e.g. from the modern Nile Delta and historically in Europe.

The semblance of 'positive symbiosis' between men and animal needs careful interpretation however. It is often confounded with the effect of fertile soils and sufficient water that helps to produce more biomass than pastoral systems per unit area, equally benefitting man and animals. The symbiosis can also be based on imported feed, either from grazing lands, e.g. outfields, or from other crop producing systems. Livestock then convert biomass from outside the system into dung and draught for crops within the system: the so-called 'Konzentrazionswirtschaft' i.e. concentration culture (McCourt, 1955; Willerding, 1980) It is a form of mixed farming that exploits the 'outfields' in favour of the 'infields', possible only in the expansion and the HEIA mode.

The numbers of humans and animals can correlate positively in systems with relative abundance of land, especially where draught, dung production and wealth accumulation by livestock is important. Some studies show increased numbers of animals per area unit as farm size declines, but there is evidence that the correlation becomes negative when pressure on land increases beyond a threshold value. The outfield / infield ratio becomes too small in such cases (De Lasson, 1981; Vaidyanathan, 1988; George *et al.*, 1989), as described by Jackson (1983):

> '*cattle numbers have started to decline in Kerala, West Bengal and Bangladesh'. [...] 'the phase of decline is marked by a high proportion of cultivated land to uncultivated land,*

and/or the extensive degradation/extreme subdivision of cultivated land. The individual
family no longer has enough feed to maintain the ideal component of livestock'.
Chayanov signals a declining outfield / infield ratio in Russian conditions of the early 20th
century as mentioned in Ch. 1. The same principle is reported for Western Europe
throughout the middle ages by Lord Ernle (1961) and Slicher Van Bath (1963). As a
combined effect of (fire-)wood collection, cropping and grazing, entire villages in Europe
disappeared in the so-called 'Wüstungen' (deserts) of Germany as well as in the Netherlands
and elsewhere (Heidinga, 1987; Castel *et al.*, 1989). For those who now travel the
intensively cropped and deforested Gangetic plains, the long-term change from using forest
and waste land grazing to crop biomass for feed are clear from the quote by Randhawa
(1980):

> *the Vedic Aryans were primarily pastoral. When they settled in the Punjab, they cut the*
> *jungles, and built their villages. They grazed their cattle in the jungles, and planted barley*
> *in the land close to the habitation where it could be protected from wild animals.*

In many densely populated areas, one cannot escape the impression that, in the absence of
external feed resources, including those from common grazing lands, livestock numbers per
farm system decrease, despite not always reliable government statistics that suggest the
opposite. Short-term effects, including temporary feed imports, disease outbreak or
droughts may conceal long-term developments. Ifar *et al.* (in preparation) showed for
villages on East Java, that more feed is now available and more animals are now kept than
10-20 years ago, but they also acknowledge that part of that feed comes from outside the
regional farm system. Ibrahim *et al.* (1991) report increases of large and small livestock
numbers on Java in the Malang and Pasuruan regencies between 1980 and 1989, probably
due to increased imports of feed or exploitation of hitherto unutilized grazing grounds.
Petheram (1986, table 1.1) reports on the other hand, that numbers of large cattle on Java
remain static or decline where numbers of small animals and poultry increased from 1969-
1983!

FARMING SYSTEMS CLASSIFICATION AND CRITERIA FOR SUCCESS

A final word is needed on the changes of systems in time and space, and on the effect of
this on the criteria for system success. As systems tend to maximize the output from their
limiting factors (Marten, 1988; Spedding, 1988) it follows that the criteria for system
evaluation must change with shifts in the resource / demand patterns, again a principle well
known in folk wisdom, e.g. *dance according to the tune.*

Change of criteria, and development of new criteria is discussed for economics by Galbraith
(1986), for project planning by Lutz (1993) and Van Pelt (1993), for scientific paradigms by
Harman (1994), and for ecology by Odum (1971). Extension/ development priorities for
livestock development also change between farmers in scales of time and space (Gahlot *et*
al., 1993). In fact, it can be assumed that men's perception of the gods change as systems
adjust to resource / demand patterns, a notion not foreign to materialistic anthropology,
cultural ecology or psychology (Geertz, 1963; Odum, 1971; Harris, 1988; Baring, 1994).

Not one, but a set of criteria may be required for evaluation of system success, and trade-offs occur when not one, but more resources become limiting, or when not only the success of one subsystem alone is measured (Conway and Barbier, 1990). Not only physical criteria, but also socio-economic aspects are important, including those of equitability (Behnke, 1985; Conway, 1985; Marten, 1988) The attempts to take into account more than one criteria for system success represent a move from reductionism towards holism. A recurring theme of this thesis is that adoption of a farming system approach implies that, depending on the mode of farming, one should not consider criteria such as only grain yield, liveweight gain or milk production to assess the production of a system. Rather, a combination of such criteria can be required. Fortunately, plant breeders are starting to appreciate the point that a crop is more than yield alone (Nordblom and Halimeh, 1982; Traxler and Byerlee, 1993: Joshi *et al.*, 1994). Many animal development officers, more than practical farmers, still have to grasp that high-milk yielders are not always the best to improve resource utilization. The thought experiments reported in Ch. 5.2 stress that in closed systems, an increased total system output may require less than maximum output of individual crop or livestock subsystems. Implicitly, this indicates the need for adjustment of one subsystem's criteria for success to the well-being of other subsystems, i.e. attention to issues of equity between farm systems (Conway, 1985), a concept that can be at odds with the approach that stresses development of individual farms.

The issue of adjusted criteria is the core, need and problem of FSR&E and system classifications. De Boer (1985) says:

> *formulation and execution of agricultural policy based on FSR&E is handicapped by its micro nature - at this level farming systems diversity becomes apparent and the researcher has difficulty coming up with general economic or agricultural policies that consistently produce the desired effect. Policymakers, on the other hand, desire policies that can be implemented with available instruments at the national or regional level. They don't like to hear the FSR&E specialist's plea that every farm is different, that government policies are contradictory or have no effect on the small farmer, or that policies may have to be tailored for very specific regions or production systems and implemented at the local level.*

If the criteria for development, and even men's perception of the gods changes with shifting resource / demand patterns, the 'near-religious' pursuit of high individual production levels in the HEIA mode of the 'developed' world needs to be reconsidered. Also, the sacred cows of the Indians, and the revered pigs of the Papuas, will become a source of contention with changing feed biomass availability and the increased demand for meat and milk. Ultimately, one might hope that the politicians' and donor's 'religion' of simple-to-transfer and universally-applicable quick fixes is replaced with the 'common sense' of niche solutions for niche problems, a reason for system classification.

CONCLUDING COMMENTS

Systems change in time and space due to shifting resource / demand patterns. They respond amongst others, by expansion, by application of innovation or by adjustment of demand to resources. Development in this context - i.e. the introduction of technological and management innovations - is not necessarily a sign of progress, but it is often driven by increased need.

A two-dimensional matrix has been designed by using existing concepts, and based on the premise that farming systems change according to shifting resource / demand patterns. Access to resources is reflected in changes of relative availability of the classical production factors, land, labour and capital on the vertical axis. A refinement in terms of the land use for livestock and/of crops, is given on the horizontal axis. The classification serves in the following chapters to discuss the role of straw as livestock feed in various farming systems.

An implicit classification criterium underlying the matrix, refers to whether a system can be considered to be of a closed or open nature, here equated with LEIA on the one hand, and expansion and HEIA on the other hand. The degree of openness determines the extent to which the demand has to be adjusted to the resources or vice versa, the basis of discussions in Ch. 2.3, 5.1, 5.2 and 6.

Criteria for system success need to reflect the limiting factors of the system under consideration. Ideally, in an holistic approach, the criteria for one subsystem take into account the criteria and wellbeing of the other subsystems. Blanket criteria, i.e. the use of standard criteria for all sorts of farms, is therefore misleading in variable systems. The role and importance of straw as animal feed varies between systems, but it is highest in the central column of the matrix, i.e. mixed crop-livestock systems. Mixed crop-livestock systems themselves exist on a scale from diversification to integration. The discussion on methods to feed crop residues as feed in the different mode of mixed crop-livestock systems, is the topic of the following chapter.

ACKNOWLEDGEMENT

Special thanks are due to: G.J.M. Oomen for discussing aspects of crop-livestock integration, L.'t Mannetje for suggesting a matrix of systems, and A.J. De Boer for stressing the importance of induced innovations, and A. Maas for preparing the matrix. Also thanks to P. Hoekstra, B.H.Slicher van Bath and G. Montsma for their suggestions and reading of an earlier draft.

REFERENCES

Achterhuis, H., 1988. Het rijk van de schaarste; van Thomas Hobbes to Michel Foucault. Ambo, Baarn, The Netherlands. 365 pp.

Altieri, M., 1991. How best can we use biodiversity in agroecosystems? Outlook on Agriculture, 20(1): 15-23.

Amartya Sen, 1981. Poverty and famines. An essay on entitlement and deprivation. ILO, Oxford University Press, New York.

Anon., 1981. Landruil. Stichting Proefboerderijen IJsselmeerpolders, subcommissie 'Landruil'. Rapport nr. 23, december 1981. 16 pp. + 13 annexes.

Bakker, H., Zemmelink, G., Uwland, J., Van Den Berg, J.C.T., Kortenhorst, L.F., Lenselink, A.J. and Knol, J., 1982. Feasibility Study dairy development East Java, Indonesia (03/10/1981 - 04/12/1981). Final Report. Dept. Tropical Animal Production, Agricultural University Wageningen, The Netherlands.

Baring, A., 1994, Towards a new Image of the Divine, Metanoia, Spring 1994, p.31-36.

Bartholomew, P.W., Ly, R., Nantoume, H., Kone, N'G., Traore, B., Khibe, T., Sissoko, K. and Baur, H., 1992. Dry-season cattle fattening by smallholder farmers in the semi-arid zone of Mali. Afr. Livest. Res., 2: 31-38.

Barton D., 1987. Draught Animal Power in Bangladesh. Ph.D. Thesis, School of Development Studies of the University of East Anglia, Norwich.

Bayliss-Smith, T.P., 1991. The ecology of Agricultural systems. Cambridge University press, Cambridge. 112 pp.

Behnke, R.H., 1985. Measuring the Benefits of Subsistence *versus* Commercial Livestock Production in Africa. Agric. Systems, 109-135.

Bieleman, J., 1987. Boeren op het Drentse zand 1600-1910. Een nieuwe visie op de 'oude' landbouw. Ph.D. Thesis, Wageningen / Utrecht, The Netherlands. 834 pp.

Bieleman, J., 1990. De verscheidenheid van de landbouw op de Nederlandse zandgronden tijdens 'de lange zestiende eeuw'. BMGN, 105(4): 527-552.

Bieleman, J., 1992. Geschiedenis van de landbouw in Nederland, 1500-1950: veranderingen en verscheidenheid. Boom, Meppel, 423 pp.

Biewinga, E.E., Aarts, H.F.M., and Donker, R.A., 1992. Melkveehouderij bij stringente normen, bedrijfs- en onderzoeksplan van het Proefbedrijf voor Melkveehouderij en Milieu. De Marke, Hengelo, Rapport No. 1. 284 pp.

Boserup, E., 1965. The conditions of agricultural growth, the economics of agrarian change under population pressure. George Allen & Unwin Ltd., London. 124 pp.

Bromley, D., (ed.), 1992. Making the commons work: theory, practice and policy. Institute for Contemporary studies. San Francisco. 339 pp.

Brouwer, F.M., Thomas, A.J. and Chadwick, M.J., 1991. Land use changes Europe. Kluwer Academic Publishers. 529 pp.

Bryant, A.M., 1986. Major determinants of milk production from grazed pasture in New Zealand. p. 3-6. In: Int. Dairy Fed. Bull. No. 199, Auckland, New Zealand.

Budelman, A., and Van Der Pol, F., 1992. Farming system Research and the quest for a sustainable agriculture. Agroforestry Systems, 19: 187-206.

Byerlee, D., Iqbal, M., and Fischer, K.S., 1989. Quantifying and valuing the joint production of grain and fodder from maize fields: evidence from Nothern Pakistan. Expl. Agric., 25: 435-445.

Campbell, B.M.S. and Overton, M. (eds.), 1991. Land, Labour and Livestock: historical studies in European agricultural productivity. Manchester University Press, Manchester and New York. 500 pp.

Cancian, F. 1989. Economic Behaviour in Peasant Communities. Ch. 6. p.127-170. In: S. Plattner (ed.), Economic Anthropology. Stanford University Press. 487 pp.

Castel, I., Koster, E., and Slotboom, R., 1989. Norphogenetic aspects and age of late Holocene eolian drift sand in Northwest Europe. In: Zeitschrift für Geomorphologie NF. 33(1): 1-26.

Conway, G.R., 1985. Agroecosystem analysis. Agric. Admin., 20: 31-55.

Conway, G.R. and E.B. Barbier, 1990. After the Green Revolution, Sustainable agriculture for Development, Earthscan Publications Ltd. London. 205 pp.

Cox, G.W., and Atkins, M.D., 1974. Agricultural ecology, an analysis of world food production systems. W.H. Freeman and Company, San Francisco. 721 pp.

Crosby, A.W. 1986, Ecological Imperialism, the Biological Expansion of Europe (900-1900). Cambridge University Press. Cambridge, U.K. 368 pp.

Crotty, R., 1980. Cattle, Economics and Development. Commonwealth Agricultural Bureaux, Farnham Royal, Slough, 253 pp.

De Boer, A.J., 1985. Some basic features of Asian livestock-production systems factors underlying productivity-improvement programs. Winrock International Livestock Research and Training Centre, Petit Jean Mountain, Morrilton, Arkansas.

De Lasson, A., 1981. Socio-economic aspects of livestock keeping in Noakhali district, Bangladesh. p. 456-495. In: M.G. Jackson, F. Dolberg, C.H. Davis, M. Haque and M. Saadullah (eds), Maximum Livestock Production from Minimum Land. Bangladesh Agricultural University, Mymensingh.

De Vries, E., 1947. Problemen van de Javaanse landbouw. Inaugurele rede 13 juni 1947. 15 pp.

De Wit, C.T. 1969, Een bodemvruchtbaarheids theorie uit de eerste helft van de 19e eeuw, Landbouwkundig Tijdschrift, 81(8): 245-251.

De Wit, C.T., Huisman, H. and Rabbinge, R., 1987. Agriculture and its environment: are there other ways? Agric. Systems, 23(3): 211-236.

DGIS, 1987. Assisting Livestock Development, Experience of development Cooperation with reference to livestock in the period 1978 - 1984. Evaluation Report Netherlands Development Cooperation, The Hague, 80 pp.

Duckham, A.N. and Masefield, G.B., 1970. Farming systems of the world, Chatto & Windus, London, 542 pp.

Ellul, J., 1990. The technological bluff. Translated by G.W. Bromley. William B. Eerdmans Publishing Companay, Grand Rapids, Michigan. 418 pp.

FAO, 1977, New Feed Resources, Proc. of a Technical Consultation, Rome, 22-24 November 1976, FAO, Rome, 300 pp.

FAO, 1991. FAO expert consultation on legume trees and other fodder trees as protein sources for livestock. Report, Kuala Lumpur, Malaysia, 14-18 October 1991, FAO, Rome. 45 pp.

Fisher, G., 1978. Farm life and its changes. Pequea Publishers. Gordonville, Pa, USA. 384 pp.

Francis, C.A., Flora, C.B. and King, L.D. (eds.), 1990. Sustainable agriculture in temperate zones. John Wiley & Sons, New York, 487 pp.

Fresco, L.O., 1986. Cassava in shifting cultivation, A systems approach to agricultural technology development in Africa. Ph.D. Thesis, Agricultural University Wageningen, Royal Tropical Institute, Amsterdam, 240 pp.

Fresco, L.O., and Westphal, E., 1988. A hierarchical classification of farm systems. Expl. Agric., 24: 399-419.

Galbraith, J.K., 1987. Economics in perspective, a critical history. Houghton Mifflin Company, Boston, 324 pp.

Gahlot, O.P., Rao, S.V.N, Pradhan, P.K., Desai, S.M., Shukla, S.G. and Jape, A.S., 1993. Felt needs of dairy farmers. A case study from cattle development centres in Saurashtra, Gujarat. p. 461-469. In: Kiran Singh and J.B. Schiere (eds.), Feeding of Ruminants on Fibrous Crop Residues: Aspects of Treatment, Feeding, Nutrient Evaluation, Research and Extension. Proc. of an Int. Workshop, 4-8 February 1991, National Dairy Research Institute, Karnal (Haryana-India), Indian Council for Agricultural Research, Krishi Bhawa, New Delhi, 486 pp.

Gass, G.M. and Sumberg, J.E., 1993. Intensification of livestock production in Africa: experiences and issues. Draft. Overseas Development Group, School of Development Studies, University of East Anglia, Norwich NR4 7TJ.

Geertz C., 1963. Agricultural involution, the process of ecological change in Indonesia. Published for the Association of Asian Studies by University of California Press, Berkeley/ Los Angeles/London. 176 pp.

George, P.S., Nair, K.N., Strebel, B., Unnithan, N.R. and Wälty, S., 1989. Policy Options for Cattle Development in Kerala. Submitted to the Kerala Livestock Development Board, Trivadrum, India and Intercooperation / Swiss Development Cooperation, Bern, Switzerland. 84 pp.

George, S., 1990. Agropastoral Equations in India: Intensification and Change of Mixed Farming Systems. p. 119-143. In: J.G. Galaty and D.L. Johnson (eds.), The World of Pastoralism; Herding Systems in Comparative Perspective. Guilford Press, New York, London. 436 pp.

Gever, J., Kaufmann, R., Skole, D. and Vorosmarty, C., 1986. Beyond oil; The threat to food and fuel in the coming decades. Complex Systems Research Centre Univ. of New Hampshire, Ballinger Publishing Company, Cambridge, Massachusetts. 304 pp.

Gibson, T., 1987. A ley farming system using dairy cattle in the infertile uplands, Northeast Thailand. World Anim. Rev., 61: 36-43.

Grigg, D.B. 1974. The agricultural systems of the world, An Evolutionary Approach. Cambridge University Press, Cambridge, 358 pp.

Grigg, D. 1982. The dynamics of agricultural change, the historical experience. Hutchinson, London. 260 pp.

Grijseels, G., 1988. Role of livestock on mixed smallholder farm in the Ethiopian highlands. A case study from the Baso and Worena werada near Debre Berhan. Ph.D. Thesis, Agricultural University Wageningen, Dept. Tropical Animal Production, Wageningen. 249 pp.

Harman, W., 1994, Cause for Change, Cause for Hope? Metanoia, Spring 1994, p. 23-31.

Harris, M., 1974. Cows, Pigs, Wars & Witches. The Riddles of Culture. Random House, New York.

Harris, M., 1985. Good to Eat. Riddles of Food and Culture. Simon and Schuster, New York.

Harris, M., 1988. Culture, People, Nature: an Introduction to General Anthropology. 5th edn. Harper and Row, New York. 678 pp.

Hayami, Y. and Ruttan, V.W., 1985. Agricultural Development, An International Perspective. The Johns Hopkins University Press, Baltimore and London. 506 pp.

Hecht, S.B., 1993. The logics of livestock and deforestation in Amazonia: directly unproductive, but profitable investments (DUPs) and development. In: Proc. VII World Conference on Animal Production, Edmonton, Alberta, Canada, 28 June-2 July 1993, Vol. 1, Invited papers.

Heidinga, H.A., 1987. Medieval settlement and economy north of the lower Rhine. Archeology and history of Kootwijk and the Veluwe (the Netherlands). Reeks Cingula 9. Universiteit van Amsterdam, Van Gorcum.

Hanley, N. and Lindgard, J., 1987. Controlling straw burning: farm management modelling of the policy options using linear programming. J. Agric. Econ., 38(1): 15-26.

Herlemann, H.H., 1954. Technisierungsstufen der Landwirtschaft, Versuch einer Erweiterung der Intensitätslehre Thünens. Zeitschrift für Agrarpolitik und Landwirtschaft, 32: 335-342.

Ibrahim, M.N.M, Wassink, G.J. and De Jong, R., 1991. Dairy Development in the province of East Java, Indonesia, 1980 - 1990. Department of Tropical Animal Production, Agricultural University Wageningen, 104 pp.

Ifar, S., Solichin, A.W., Udo, H.M.J. and Zemmelink, G., 1995. Subsistence farmers' access to cattle via sharing in the limestone area in the Southern part of East Java, Indonesia, (in print).

Iniguez, L.C. and Sanchez, M.D. (eds.), 1990. Integrated tree cropping and small ruminant production systems. Proc. of a Workshop on Research Methodologies. Medan, North Sumatra, Indonesia, 9-14 September 1990. Sponsored by Agency for Agricultural Research and Development, Small Ruminant-Collaborative Research Program, International Development Research Centre. 329 pp.

Jackson, M.G., 1983. A strategy for improving the productivity of livestock in the hills of Uttar Pradesh. p.130-154. Proc. of the Seminar: 'Regeneration of the Himalayas: Concepts and Strategies', 24-25 October 1983, Nainital, UP, India.

Jahnke, H.E., 1982. Livestock Production Systems and Livestock Development in Tropical Africa. Kieler Wissenschaftsverlag Vauk, Germany. 253 pp.

Joshi, A.L., Doyle, P.T. and Oosting, S.J., 1994. Variation in the Quantity and Quality of Fibrous Crop Residues. Proc. of a Nat. Seminar, BAIF Development Research Foundation, Pune (Maharashtra-India), 8-9 February 1994, Pune, India, 174 pp.

Kang, B.T., Reynolds, L. and Atta-Krah, A.N., 1990. Alley farming. Advances in Agronomy, 43: 316-359.

Kaplan, R.D., 1994. The coming anarchy. The Atlantic Monthly, February: 44-76.

Kelley, J., 1992. Upgrading of waste cereal straws. Outlook on Agriculture, 21 (2): 105-108.

Kiran Singh and Schiere, J.B., (eds.), 1993. Feeding of Ruminants on Fibrous Crop Residues. Aspects of Treatment, Feeding, Nutrient Evaluation, Research and Extension. Proc. of an Int. Workshop, 4-8 February 1991, National Dairy Research Institute, Karnal (Haryana-India), Indian Council for Agricultural Research, Krishi Bhawa, New Delhi, India, 486 pp.

Kraybill, D.B., 1993. The Riddle of Amish Culture. The Johns Hopkins University Press, Baltimore and London.

Lagemann, J., 1977. Traditional African Farming Systems in Eastern Nigeria; An analysis of Reaction to Increasing Population Pressure. Weltforum Verlag Munchen, (Afrika-Studien: Nr. 98).

Longman, 1985. Longman Concise English Dictionary. Longman, Avon, U.K. 1651 pp.

Lord Ernle, 1961, English Farming, Past and Present, 6th edition, Heinemann, London / Melbourne / Toronto, Frank Cass & Co, London. 559 pp.

Lutz E., 1993. Towards improved accounting for the environment. Worldbank quoted from The Economist, July 31. 63 pp.

't Mannetje, L., 1989. Tropical grassland improvement. Lecture Notes. Dept. of Field Crops and Grassland Science. Agricultural University Wageningen, 39 pp.

't Mannetje, L., and Jones, R.M., 1992. Plant resources of South-East Asia, No. 4: Forages. PROSEA, Bogor Indonesia. 300 pp.

Marten, G.G., 1988. Productivity, stability, sustainability, equitability and autonomy as properties for agroecosystem assessment. Agric. Systems, 26: 291-316.

Martin, W.S., 1944. Grass covers in their relation to soil structure. Empire J. Expt. Agri., 21-33.

McCourt, D., 1955. Infield and outfield in Ireland. Ec. Hist. Rev., Second Series, 7: 276-369.

Meadows, D.H., Meadows, D.L. and Randers, J., 1972. The Limits to Growth: a Report for the Club of Rome's Project on the Predicament of Mankind. New York, 207 pp.

Meertens, H.C.C., Ndege, L.J. and Enserink, H.J., 1994. Farming Sytems Dynamics, Changes in Time and Space in Sukumaland, Tanzania. KIT-publication (manuscript), Royal Tropical Institute, Amsterdam, (in press).

Montsma, G., 1984. Introduction to Animal Farming Systems, Lecture notes, Dept. Tropical Animal Production, Agricultural University Wageningen.

Nestel, B. (ed.), 1984. Development of Animal Production Systems, World Anim. Sci. A2, Elsevier, Amsterdam, 435 pp.

Netting, R.McC. 1993. Smallholders, Householders, Farm Families and the Ecology of Intensive, Sustainable Agriculture. Stanford University Press, Stanford.

Nordblom, T. and Halimeh, H., 1982. Lentil crop residues make a difference. p.8-10. In: Lentil Exp. News Service (LENS), Newsl. No.9. The International Centre for Agricultural Research in the Dry Areas (ICARDA); University of Saskatchewan, Syria. 46 pp.

Nou, J., 1967. Studies in the Development of Agricultural Economics in Europe. Almquist & Wiksells Boktryckeri AB, Uppsala.

Noy-Meir, I. and Seligman, N.G., 1979. Management of semi-arid ecosystems in Israel. p.113-160. In: B.H. Walker (ed.), Management of Semi-arid Ecosystems, Elsevier Scientific Publishing Company, Amsterdam. 398 pp.

NRC, 1981. Food, Fuel and Fertilizer from Organic Wastes. Report of an *ad-hoc* panel of the Advisory Committee on Technology Innovation Board on Science and Technology for International Development, Commission on International Relations, National Research Council, Washington, DC.

NRC, 1989. Alternative Agriculture. Committee on the role of alternative farming methods in modern production agriculture. Board on agriculture National Research Council, National Academy Press, Washington, DC, 448 pp.

NRC 1991. Microlivestock, Little-known Small Animals with a Promising Economic Future. National Academy Press, Washington, DC.

Odum H.T., 1971. Environment, Power and Society. Wiley Interscience, New York, 331 pp.

Palthe, J.G.L., 1989. Upland farming on Java, Indonesia: a socio-economic study of upland agriculture and subsistence under population pressure. Netherlands Geographical Studies 97, Amsterdam / Utrecht.

Patil, B.R., Rangnekar, D.V. and Schiere J.B., 1993. Modelling of crop-livestock integration: effect of choice of animals on cropping pattern. p. 336-343. In: Kiran Singh and J.B. Schiere (eds.), Feeding of Ruminants on Fibrous Crop Residues: Aspects of Treatment, Feeding, Nutrient Evaluation, Research and Extension. Proc. of an Int. Workshop, 4-8 February 1991, National Dairy Research Institute, Karnal (Haryana-India), Indian Council for Agricultural Research, Krishi Bhawa, New Delhi, India, 486 pp.

Pearson, C.J. (ed.), 1992. Ecosystems of the World - Field Crop Ecosystems. Elsevier, Amsterdam, London, New York, Tokyo, 570 pp.

Pingali, P., Bigot, Y. and Binswanger, H.P., 1987. Agricultural Mechanization and the Evolution of Farming Systems in Sub-Saharan Africa. World Bank, The Johns Hopkins University Press, Baltimore and London. 216 pp.

Poelhekke, F.G.M.N., 1984. Prikkeldraad in het oerwoud. De veeteelt in het process van economische en sociale integratie van het Amazonegebied in Brazilië. Geografic and Planologic Institute, Catholic University Nijmegen, Nijmegen, 371 pp.

Ponting, C., 1991. A Green History of the World. Penguin Books, London. 432 pp.

Posner J.L. and Gilbert E., 1991. Sustainable agriculture and farming systems research teams in semiarid West Africa: a fatal attraction. J. Farming System Res. Ext., 2(1): 71-86.

Powell, J.M. and Waters-Bayer, A., 1985. Interactions between livestock husbandry and cropping in a West African savanna. In: J.C. Tothill and J.J. Mott (eds.), Ecology and Management of the World's Savannas. CSIRO, Division of Tropical Crops and Pastures, Queensland, Australia. The Australian Academy of Science, Canberra / C.A.B., Farnham Royal, Buckinghamshire.

Preston, T. and Murgueitio, E., 1992. Strategy for Sustainable Livestock Production in the Tropics. CIPAV, AA20591, Cali, Colombia. 89 pp.

Ramírez, R.G., Garza, J., Martínez, J. and Ayala, N., 1991, Wood ash, sodium hydroxide and urine to increase sorghum straw utilization by sheep. Small Rum. Res., 5: 83-92.

Randhawa, M.S., 1980. A History of Agriculture in India. Volume I. Beginning to 12th century. Indian Council of Agricultural Research, New Delhi. 541 pp.

Reed, J.D., Capper, B.S. and Neate, P.J.H. (eds.), 1988. Plant Breeding and the Nutritive Value of Crop Residues. Proc. of a Workshop, ILCA, Addis Ababa, Ethiopia, 7-10 December 1987. ILCA, Addis Ababa, Ethiopia, 334 pp.

Reijntjes, C., Haverkort, B. and Waters-Bayer, A., 1992. Farming for the Future: An Introduction to Low-External-Input and Sustainable Agriculture. Macmillan, ILEIA. 250 pp.

Reynolds, S.G., 1980. Grazing cattle under coconuts. World Anim. Rev., 35: 40-45.

Rifkin, J., 1989. Entropy into the Greenhouse World. Bantam Books, New York, revised edn. 354 pp.

Rifkin, J., 1992. Beyond Beef: the Rise and Fall of the Cattle Culture. A Dutton book, Penguin, New York, 353 pp.

Roep, D. and De Bruin, R., 1994. Regional marginalization: styles of farming and technology development. p. 217-227 In: J.D. Van Der Ploeg and A. Long (eds.), 1994. Born from within: practice and perspectives of endogenous rural development. Van Gorcum, Assen. 298 pp.

Röling, N. 1989. The Agricultural Research-technology Transfer Interface: a Knowledge Systems Perspective. International Service for National Agricultural Research (ISNAR). The Hague.

Ruthenberg, H., 1980. Farming Systems in the Tropics. 3rd edn. Clarendon Press, Oxford, 424 pp.

Schiere, J.B. and Ibrahim, M.N.M., 1989. Feeding of Urea-ammonia Treated Rice Straw: A Compilation of Miscellaneous Reports Produced by the Straw Utilization Project (Sri Lanka). Pudoc, Wageningen, 125 pp.

Schiere, J.B., 1992. Veeteelt en bodemvruchtbaarheid, concentreren of conserveren? In: Verslag van de vierenzestigste tropische landbouwdag; Nutriëntenstromen en duurzaam landgebruik in de tropen. 11 november 1992, Nili-KGvl, Wageningen en KIT, Amsterdam.

Simmons, I.G., 1989. Changing the Face of the Earth; Culture, Environment, History. Basil Blackwell, New York / Oxford. 487 pp.

Simpson J.R., 1988. The economics of livestock systems in developing countries, Westview Special Studies in Agriculture Science and Policy. 297 pp.

Slicher Van Bath, B. H., 1963. The Agrarian History of Western Europe, A.D. 500-1850. Translated by O. Ordish. Arnold, London. 364 pp.

Smith, J., Barau, A.D., Goldman, A. and Mareck, J.H, 1993. The role of technology in Agricultural Intensification: The evolution of maize production in the Northern Guinea Savannah of Nigeria. p. 144-166. In: K.A. Dvorák (ed.), Social Science Research for Agricultural Technology Development, Spatial and Temporal Dimensions. Proc. of IITA-workshop 2-5 October 1990, CAB-International.

Spedding, C.R.W., 1988. An Introduction to Agriculture Systems. 2nd Edition, Elsevier Applied Science, London. 189 pp.

Stocking, M and Abel N. 1981. Ecological and environmental indications for the rapid appraisal of natural resources. Agric. Admin., 8: 473-484.

Sumberg, J.E. and Atta-Krah, A.N., 1988. The potential of alley farming in humid West Africa - a re-evaluation. Agroforestry Systems, 6: 163-168.

Sundstøl, F. and Owen, E. (eds.), 1984. Straw and Other Fibrous By-products as Feed. Developments in Anim. Veterinary Sci., 14. Elsevier. 604 pp.

Tillman, A.D., 1981. Animal agriculture in Indonesia. Winrock International Livestock Research and training Centre, Arkansas, 80 pp.

Traub, J.F. and Wozniakowski, H., 1994. Breaking Intractability, Scientific American, January: 90-93.

Traxler, G. and Byerlee D., 1993, A joint-product analysis of the adoption of modern cereal varieties in developing countries. American J. Agric. Econ., 75 (November): 981-989.

Vaidyanathan, A., 1988. Bovine Economy in India. Centre for Development studies, Monograph series, Trivandrum. 209 pp.

Van Den Dries, A. and Portela, J., 1994. Revitalization of farmer managed irrigation systems in Trás-os-Montes. p. 71-100. In: J.D. Van Der Ploeg and A. Long (eds.), Born from within. Practice and Perspective of Endogenous Rural Development. Van Gorcum, Assen, 298 pp.

Van Der Ploeg, J.D. and Long, A. (eds.), 1994. Born from Within: Practice and Perspectives of Endogenous Rural Development. Van Gorcum, Assen, 298 pp.

Van Der Pol, F., 1992. Soil Mining: an Unseen Contribution to Farm Income in Southern Mali. Bull. 325, Royal Tropical Institute, Amsterdam, 48 pp.

Van Pelt, M.J.F., 1993. Sustainability-oriented Project Appraisal for Developing Countries. Ph.D. Thesis, Wageningen Agricultural University, Wageningen.

Walshe, M.J., Grindle, J., Nell, A. and Bachmann, M., 1991. Dairy development in Sub-Saharan Africa, A study of Issues and Options. World Bank Techn. Pap. No. 135, Africa Technical Department Series, The World Bank, Washington, DC, 94 pp.

Warren Wilson, J., 1972. Introduction 2: Control of crop processes. In: A.R. Rees, K.E. Cockshull, D.W. Hand and R.G. Hurd, Crop Processes in Controlled Environment. Academic Press, London.

Waters-Bayer, A. and Bayer, W., 1987. Building on traditional resource use by cattlekeepers in central Nigeria. ILEIA Newsletter, 3(4): 7-9.

Watson, G.A., 1983. Development of mixed tree and food crop systems in the humid tropics: a response to population pressure and de-forestation. Expl. Agric., 19: 311-332.

WCED, 1987. Our Common Future. Report of the World Commission on Environment and Development, Oxford University Press, Oxford, U.K. 400 pp.

Westgaard, P., and Sundstøl, F., 1986. History of straw treatment in Europe. p. 155-163. In: M.N.M. Ibrahim and J.B. Schiere (eds.), Rice Straw and Related Feeds in Ruminant Rations: Proc. of an Int. Workshop, 24-28 March 1986, Kandy, Sri Lanka, SUP Publication No.2., Dept. Tropical Animal Production, Agricultural University Wageningen, The Netherlands. 407 pp.

White, B., 1976. Population, involution and employment in rural Java. Development and Change, 7: 267-290.

Wilkinson, R.G., 1973. Poverty and Progress: An ecological Model of Economic Development. Methuen, London. 225 pp.

Willerding, U, 1980, Anbaufrüchte de Eisenzeit und des fruehen Mittelalters, ihre Anbau Formen, Standoertsverhaeltenisse und Erntemethoden, p. 126-196. In: H. Bech, D. Denecke and H. Jankuhn (eds.), Untersuchungen zur eizenzeitlichen und fruehmittelalterlichen Flur in Mitteleuropa und ihrer Nutzung, Bericht ueber die Kolloquien der Kommision fuer die Alterumskunde Mittel- und Nordeuropas in den Jahren 1975 und 1976. 2 Tl. Van Der Hoek & Ruprecht, Goettingen.

Wilson, R.T., 1986. Central Mali - Cattle husbandry in the agropastoral system. World Anim. Rev., 58: 23-30.

WRR, 1992. Grond voor keuzen. Vier perspectieven voor de landelijke gebieden in de Europese Gemeenschap. Rapporten aan de Regering, Wetenschappelijke Raad voor het Regeringsbeleid, Nr.42. SdU uitgeverij, 's-Gravenhage, 149 pp.

Chapter 2.3

LIVESTOCK AND FARMING SYSTEMS RESEARCH
III. DIFFERENT WAYS OF FEEDING CROP RESIDUES

J.B. SCHIERE and J. DE WIT

SUMMARY

The method and purpose of feeding livestock depends on the resource / demand patterns which prevail in and around the farm system. In spite of large variation between systems, it is possible to discern trends in the use of crop residues for animal feed. This paper reviews types and availability of feed biomass, functions of animals (demand), and ways in which the fibrous crops residues (straws) are fed. In conditions with relatively abundant land (expansion agriculture), livestock obtain nutrients by grazing on non-agricultural land, and crop residues are not very important as feed. When access to land and inputs is limited, e.g. in low external input agriculture (LEIA), crop residues are important as feed, as well as for other purposes such as fuel, thatching and soil conservation. In high external input agriculture (HEIA) where land and labour are relatively scarce - animals tend to be fed with fertilized fodder and imported concentrates. Straw in HEIA systems has very limited value as feed, only as a source of fibre in diets based on high levels of concentrate or lush greens. In systems that aim to balance the import and use of nutrients (new conservation agriculture), animals can convert a variety of crop residues, including those from soil conservation measures, into useful products. In a general sense, feed biomass shortages leads to a need for system adjustment, i.e. to an increased competition for crop residues between farmers, animals, soil and bedding. The search for techniques to use straw as feed is a typical case of the need for induced innovations in systems with limited access to inputs.

INTRODUCTION

Case studies of crop-livestock systems undergoing change are available for densely populated areas (De Boer and Welch, 1977; Vaidyanathan, 1988; George *et al.*, 1989; Palthe, 1989), for more sparsely populated areas (Lagemann, 1977; Steinfeld, 1988) and for pastoral systems (Jahnke 1982; Van Der Graaf 1985). Attempts to analyze more generally how livestock systems change with shifting resource / demand patterns, have been made by Andreae (1980), Ruthenberg (1980), Crotty (1980) and Pingali *et al.* (1987). However, the role of straw as feed in changing crop-livestock systems has still been insufficiently reviewed. This chapter - the third in a series - therefore describes changing methods of feeding straws in different modes of mixed crop-livestock farming, based on a description of Farming Systems Research (FSR) in the first chapter, and a classification of livestock systems in the second.

CHANGING FEED RESOURCES IN MIXED SYSTEMS

Resources for crop and animal production, can be classified into land, labour and capital as explained in Ch. 2.2. 'Land' is an aggregate term that relates to the availability of plant biomass for animal feed, even though the relation between land and access to feed is not very direct, and in spite of large differences between systems. Firstly, the biomass production per unit of land differs between, for example, the Gangetic or West European deltas, and the arid Sub-Sahelian regions. Secondly, availability does not always reflect access of individual farmers to biomass for animal feed. Thirdly, differences in use of labour and capital affect the biomass output from an area unit of land. Changing crop yields, cropping patterns and straw / grain ratios, combined with the cultivation of hitherto non-agricultural lands, further complicate the relation between access to land and feed biomass.

Changing cropping patterns and access to feed biomass

Increased cropping generally results in less forest, waste- and fallowland, catchcrop or stubble grazing. Biomass production from each of these sources varies widely in terms of quantity and quality (Cox and Atkins, 1974; Winrock, 1978; Jahnke, 1982). With irrigation and fertilization in HEIA, cropland can produce more biomass than the original waste land (Powell, 1985; Steinfeld, 1988; Joshi *et al.*, 1994). New crop varieties tend to have a lower straw / grain ratio, but a relative reduction of straw biomass is often compensated by an increase in absolute terms, due to doubled or tripled grain yields, i.e. increased total biomass. Lower straw yields per harvest can also be compensated on a year-round basis, by increased cropping intensity (Joshi *et al.*, 1994).

Generally speaking, however, a land shortage implies a changed access to plant biomass for feed. A shift of feed biomass supply from forests, waste and fallow land, to crop residues or cultivated fodder is beyond doubt as the ratio of agricultural to non-agricultural land increases. Straw/grain price ratios are known to increase in some systems, confirming a relatively higher demand for crop residues, partly generated in urban markets (Janssen *et al.*, 1990; Kelley *et al.*, 1991; De Wit *et al.*, 1993). In LEIA this is likely to lead to decreased access to feed biomass for animal owners that traditionally depend on free communal grazing areas (Panayotou and Tokrisna, 1982; Jackson, 1983; Jodha, 1986; Udo *et al.*, 1990).

Fodder cultivation or the purchase of supplements, is the standard response to feed shortage in HEIA. In LEIA, however, and almost by definition, most feed comes from within the farm system. Straw feeding becomes more important and the objective of keeping livestock has to adjust itself to the feed resources available. In addition, the better crop residues such as oil seed cakes and brans, are increasingly taken from the local farm systems to HEIA systems, amongst others caused by the centralized processing of agricultural produce.

In new conservation agriculture, the feed will have to come mainly from within the farm system. In the ideal case, tree leaves or grass from conservation ridges and bunds become available (Kang *et al.*, 1990; Nitis *et al.*, 1991) resulting in better quality 'crop residues' than those produced as by-product from grain crops. Also, and ideally in new-conservation agriculture, increased on farm use of grain milling and oilseed residues is required to avoid large scale translocation and concentration of minerals, income and use of fossil energy.

Types of crop residues

In our definition, crop residues consist of all those feeds that are by-products from cropping, such as straws and products from oilseed or grain processing. But even tree leaves or grasses grown to provide shade, firewood or protection against erosion can be considered to be crop residues. They are classified in Table 1 as poor, medium and good quality feeds, qualifications that are used to avoid confusion with conventional terminology, even though a central point of this thesis is that the qualification 'good' or 'bad' is system-dependent. The classification of Table 1 uses crude protein and total digestible nutrients (CP and TDN), as described by Zemmelink (1986), and it indirectly reflects the feed intake as established in the formulae by Ketelaars and Tolkamp (1992).

Even if supplies of total feed biomass increase, due to more crop biomass production by use of fertilizer and irrigation, it is likely that the quality of feed will decline when straws replace grazing on roadside and wasteland. Straws of high yielding varieties (HYV's) do not necessarily have a lower feed quality than those of the traditional varieties (Capper, 1990; Joshi *et al.*, 1994), but straws in general have a lower nutritive value than the green feed from forest, roadsides or fallow land (Table 1). As mentioned earlier, the quality of on-farm feed biomass is likely to decline further due to centralized milling and oilseed processing which extracts valuable feed supplements from the countryside. An important exception to the rule of decreased biomass quality in LEIA and/or new conservation agriculture, is the development of crop rotations with cruciferae and legumes, e.g. the Flemish and Norfolk systems (Lord Ernle, 1961; Slicher Van Bath, 1963), a practice continuing even today in the farming systems of northern India and the Nile Delta.

FUNCTIONS OF LIVESTOCK

Animals convert solar energy that is captured in plant biomass, into products that serve human society. On a more abstract level, this transformation of energy contributes to the organization and control of society (Odum, 1971; Ch. 6). Livestock rarely perform only one function (Winrock, 1978), and animals channel energy in various forms, according to the demand, into society. A brief description of the major functions of livestock is given here to balance the emphasis on milk in the rest of this thesis, and to describe the types of demand for animal products. Indirect effects of livestock in cropping systems, such as

Table 1   A classification of crop residues according to crude protein content (CP), energy content (TDN) and CP/TDN ratio.

Crop residue type	CP%	TDN%	CP/TDN
category I: good quality			
oilseed cake	28	70	0.40
concentrate feed	15	65	0.23
legume tree leaf	24	60	0.40
category II: medium quality			
medium quality grass	12	60	0.20
rice bran	11	55	0.20
mature tropical grasses	10	55	0.18
category III: poor quality			
maize straw	6	50	0.12
rice straw	4	45	0.09

note: These values are approximations

the possibility for diversification of cropping patterns, damage to soils and crops, and strengthening of social relationships, are important but do not significantly alter the points to be made.

Socio economic functions

The socio-economic importance of livestock is illustrated by the linguistic relation between words for livestock and money, wealth or wellbeing (Box 1 in Ch. 1). At a more abstract level, wealth and/or status can be considered to represent the combined value of animals for all their physical functions. In expansion agriculture with low natural fertility and no access to fossil fuel, cattle are a precondition for cropping, since they provide manure and draught. Human labour alone cannot till enough land to provide sufficient food, e.g. in the low productive medieval European agriculture (Crotty, 1980) or at present in sub-Saharan regions (Binswanger, 1986; Berckmoes *et al.*, 1988). Livestock are also an essential source of income or saving for landless peasants, provided there is access to free roadside grass or stubble land grazing (Harris, 1965; Jodha, 1986).

In systems with more access to resources, cattle are a store of wealth that is accumulated after cropping, reported for Botswana by Steinfeld (1988), but also known in other countries. In those farming systems the use of livestock is handy or even important as a security against misharvest or other misfortune, but is not a precondition for cultivation (Bosman and Moll, 1995).

The function of livestock in the provision of marketable produce such as milk, offspring and meat, is likely to increase relative to its saving function, particularly where banking facilities are developed as an alternative to keep money (Van Der Graaf, 1985), or where increased urban incomes raise the demand for animal proteins (Alexandratos, 1988). The saving function is also likely to diminish in conditions of decreased access to feed biomass.

Where livestock is still kept as an investment in situations of decreasing feed biomass, it is probably more a hedge against inflation than a converter of plant energy into produce (Shanmugartnam *et al.*, 1992).

Food for human consumption

Vegetarian diets are possible and humans do not depend solely on foods and energy from animal origin (Spedding, 1988). It is unlikely that animals were domesticated directly for milk or meat production, but rather for ceremonial purposes (Clason, 1977; Winrock, 1978; Rifkin, 1992). In general therefore the production of food cannot be the sole argument to keep livestock, though much depends on the availability of food, e.g. roots and tubers, or rice and pulses, or on the needs of special groups such as (reproductive) women and growing children. Livestock are essential for food production, where arid and mountainous land is not suited for crops, or where animals use crop residues that are not suitable for human consumption. Higher income tends to increase the demand for food of animal origin (Crotty, 1980; Alexandratos, 1988). It makes animals valuable for cash supply through the sale of produce, sometimes at the expense of home consumption of for example milk.

Animal power

Livestock are often indirectly essential for food production. If draught power based on fossil fuel is not available, animals provide power for cultivation of poor or heavy soils where the demand for crops cannot be met by manual labour alone. Where unreliable rains require timely operations, or where rapid transport and communications are required, e.g., in the case of war, animals provide speed as an essential commodity for the survival of society. The use of animal power on good soils permits specialization, or diversion of energy into a more elaborate organization of society. The introduction of animal draught in expansion systems in Africa, allows the expansion of cropped area for food and cashcrops (Pingali *et al.*, 1987 ; Berckmoes *et al.*, 1988). In new conservation agriculture, animal power could be used to allow more timely and better land cultivation, potentially saving on fossil fuel.

The importance of energy from animal power declines when landholdings become smaller, when energy from feed biomass becomes scarce or - alternatively - when fossil fuels become cheap (De Boer and Welsch, 1977; Panayotou and Tokrisna, 1982; Jackson, 1983; Barton, 1987). The first author has seen men pulling a plow on a Javanese paddy field as long ago as 1973, as also suggested by De Lasson (1981) for Bangladesh, and as is common in parts of China (A.J. De Boer, pers. comm. 1992).

'Production' of dung

Dung production by animals can be defined as the concentration of soil fertility from communal and marginal lands (the outfields) onto small plots (the infields). In those systems, the animals do not generate but concentrate soil nutrients, incurring large losses in the process (Schiere, 1992). This process takes place in *expansion* agriculture, and it is termed 'Konzentrations-wirtschaft' by Willerding (1980) or infield / outfield system in English literature (McCourt, 1955; Slicher von Bath, 1963). Access to artificial fertilizers, produced and/or transported mainly by use of fossil fuel in HEIA, reduces the need to conserve animal excreta. For example, sheep from the Dutch moors disappeared after the introduction of fertilizer (Bieleman, 1987). In HEIA dung disposal can even become a problem, i.e. dung has a negative value. The new conservation mode of farming requires better excreta management. The use of straws for bedding to reduce losses of urine, presents an interesting issue in this respect for the allocation of scarce straw biomass for feed, fertilizer or bedding. The intensive foddercrop rotations systems such as the ley, the Flemish and the Norfolk systems use animals to permit the incorporation of crops - or trees - that fix nitrogen, mobilize phosphate or add soil organic matter (Chayanov, 1926; Lord Ernle, 1961; Kang *et al.*, 1990; Overton, 1991).

CHANGING ROLE OF STRAW IN FARMING SYSTEMS

The use of straw for animal feed is clearly most relevant in the central column of the classification in Ch. 2.2, and dependent, among others on the functions of animals and on the non-feed use of straw. The following discussion therefore, briefly reviews uses of straw in mixed systems, but does not consider the systems with predominantly livestock or crops.

Non-feed use of straw

In systems with expansion agriculture, straw has little or no direct use, unless it can be sold to systems where feed and fibre shortages occur. In LEIA however, straw is valuable for several uses at the same time: feed, roofing, fuel and bedding, to name a few. In fact, there is competition between the different uses for straw in those systems, and hardly 'a straw' is lost. In HEIA, straw is sometimes burned (Staniforth, 1982; Kelley, 1992), unless industrial activity causes a demand for straw, e.g. for products such as paper, board or mushrooms (Hartley *et al.*, 1987). Straw burning is out of the question in new conservation farming. It constitutes an energy leak, and it leads to the release of nitrogen, sulfur, carbon monoxide and even methane into the atmosphere (Schütz *et al.*, 1990). Burning also causes a loss of organic material that is potentially valuable for bedding or composting, if not for feed or other purposes. It is obvious that the use, i.e. the competition for straw increases from expansion, via LEIA to new conservation agriculture. The trend for increased competition is relaxed in HEIA, where energy subsidies for the system permit straw energy to be left unused, i.e. where straw becomes a nuisance rather than a resource.

Straw feeding methods

The different straw feeding methods are summarized in Table 2. The method of feeding and its usefulness per farming system is tentatively indicated in Table 3. The usefulness depends on the mode of farming, the access to other feeds, the type and level of desired production. Again, the criteria for successful use of straw change between systems. Much of the reasoning in Table 3 is based on a combination of anecdotal evidence from literature, from field observations and from the work compiled by Kiran Singh and Schiere (1993, 1995). The validity of the reasoning is tested in Ch. 4.1 as far as feeding systems with supplements alone or in combination with urea-treated straw are concerned. Ch. 6 verifies the usefulness of these systems from the viewpoint of system control.

Table 2   Description of different feeding systems

feeding system	description	references
- emergency feeding and survival feeding	the use of any type of feed to achieve survival of the herd or animal, if necessary at the expense of liveweight and/or (re)production	Altona, 1966; Allden, 1970; Thole *et al.*, 1993; Ch. 5.1
- catalytic supplementation	the use of small quantities of good quality feed to improve digestion and intake of a basal ration of straw or mature grass	Alexander, 1972; Preston and Leng, 1987; Ch. 3.1, 3.2, 4.1 and 4.2
- substitutional supplementation	the use of large quantities of supplement to supply sufficient nutrients for a desired level of animal output, if necessary at the expense of straw/grass intake, i.e. supplement substitutes the basal ration	Ch. 3.3, 3.4, 4.1 and 4.2
- straw treatment	use of chemical / physical methods to improve straw quality	Ch. 3.1-3.4 and 4.1
- chopping and soaking	chopping implies the reduction of feed particle size, commonly at the size of a few centimetres or more, mostly done to avoid waste of feed	De Wit *et al.*, 1993; Badurdeen *et al.*, 1994
- selective consumption	farmers and/or animals can select the good part of the feed, leaving the residue for animals of lower output or for other uses than feed	Zemmelink, 1986; Wahed *et al.*, 1990; Ch. 5.2
- stripping and thinning	the use of leaves before they mature on plants for animal feed, mostly coarse grains such as maize and millets, and the use of purposely dense sown plants for animal feed	Byerlee *et al.*, 1989; Singh and Saha, 1995
- variability	this term implies the use of differences in straw quality and quantity due to management, environment and genetic factors	Reed *et al.*, 1987; Joshi *et al.*, 1994
- adjusted cropping	a variation on the theme of variability (see above): crop choice is at least partly based on the nutrient requirements of the animals; or animals and crops are mutually adjusted, e.g. in the Flemish and the Norfolk systems (see text)	Nordblom, 1983; Patil *et al.*, 1993; Ch. 2.2

Expansion agriculture essentially has an abundance of forest, bush and waste land grazing. These feed resources provide a better source of nutrients than straw, and straws can therefore be left in the field or burned. Straw in these systems is only useful to help the animals through a period of feed scarcity. If straw is fed, it is generally possible to apply selective consumption, i.e. animals can be allowed to refuse inferior parts of the feed (Zemmelink, 1986; Wahed *et al.*, 1990; De Wit *et al.*, 1993).

The use of straw for feed is most common in the LEIA and new conservation agriculture, systems that are both characterized by an adjustment of objectives to resources (Ch. 2.2 and 5.1). Particularly in LEIA, the shortage of feed, combined with the availability of labour and the adjustment of animal output to poor quality feed resources, makes it relevant to chop or soak the straw in order to avoid wastage, or to make sure that a maximum number of animals is maintained (De Wit *et al.*, 1993; Ch. 5.2).

In HEIA, the animals are mainly fed with cultivated fodders and purchased concentrate feeds. In these systems, it generally pays to adjust the feeds to the production objective and straw has no other use than to prevent overfeeding, or to serve as a source of fibre, e.g., to prevent bloat or acidosis.

Table 3　　　Usefulness of straw feeding methods per mode of farming in mixed crop livestock systems: a first approximation

Mode of agriculture[d]	Relevant feeding systems
expansion	- emergency feeding - selective consumption - catalytic supplementation
LEIA	- emergency feeding - chopping and/or soaking to avoid wastage - stripping / thinning - variability - straw treatment with kitchen ash or urine
New conservation agriculture	- straw treatment with urea / NaOH - selective consumption - adjusted cropping - variability
HEIA	- substitutional supplementation - straw as source of fibre in high concentrate rations, e.g. against acidosis

Note:　for explanation see text and Ch. 2.2

Straw use in new conservation farming is determined by the need to better recycle or preserve excreta, to maintain soil structure and to avoid straw burning. The restrained use of inputs to utilize straw for animal feeding, e.g. ammonia treatment, is an option in these systems as shown in Norway (Westgaard and Sundstøl, 1986). The increased use of straw requires an adjustment of animal production to the resources available (Kidane, 1984; Patil *et al.*, 1993). The use of straw for competing functions, e.g. feed for different classes of animals, bedding and roofing, particularly in LEIA and new conservation agriculture, represents an interesting topic for allocation studies, e.g., is straw to be used either as animal feed, for bedding to collect urine, for the soil as mulching, for industry, roofing or fuel. (Hartley *et al.*, 1987; Lal, 1988; Budelman and Van Der Pol, 1992; Lamers and Feil, 1993). Farmers themselves have developed intricate systems where combinations of straw use are possible. For example:

straw can be first be fed to the animals, the left-overs after selective consumption are used for bedding, and the final residue is mixed with the dung for composting or dungcakes. Straw left over after selective consumption can also be fed to dry and 'unproductive' animals.

Straw use in these integrated systems must also be seen in relation to the possibility of increased on-farm recycling of grain and oilseed milling products. The extraction of these better quality feeds from LEIA farm systems, implies that the options for livestock production in these systems are reduced, i.e. that nutrients and income become translocated and concentrated into the HEIA systems. The result on sustainability of farming system, i.e. dung disposal and between-farm systems problems of equity are disquieting (Conway, 1985; Conway and Barbier, 1990; Durning and Brough, 1991; Kaasschieter *et al.*, 1992).

CONCLUSIONS

Livestock are an integral part of many farming systems, but the access to quality and quantity of feed biomass, the availability of energy from other sources and the demand for animal produce determine the functions of animals and the need for crop residues as feed. Access to feed biomass is affected by decreased availability of wasteland grazing and more intensive use of land for cropping. Straw has hardly any role as a feed in expansion agriculture. LEIA systems attempt to compensate feed shortages by increased use of straws and by adjusting the function of animals to the feed supply. HEIA uses inputs such as concentrates or fertilizer for fodder cultivation, and straws have virtually no function as an animal feed. New conservation agriculture uses straw for feed to some extent, but also for other purposes, e.g. bedding and for recycling of nutrients on farm. Even though fibrous crop residues can be an important source of feed biomass, particularly in LEIA and new conservation agriculture, the use of straws is limited by their low nutritive value. Changing biomass availability and the demand for animal products therefore requires methods to improve straw utilization for animal feed. It appears that the usefulness of these methods can be indicated per mode of farming. Competition between livestock, soil and other parts of the farming system for organic matter, as well as equity in the allocation of resources, are relevant issues for further research. They emerge directly from the changing and often decreased relative access of an increasing human population, to energy from plant biomass.

ACKNOWLEDGEMENT

Thanks are due to G.J.M. Oomen whose insight in soil - plant - animal relationships has provided us with many useful suggestions.

REFERENCES

Alexander, G.I., 1972. Non Protein Nitrogen Supplements for Grazing Animals in Australia. World Anim. Rev., 4 (11): 11-14.

Alexandratos, N., (ed.), 1988. World Agriculture toward 2000. An FAO study. FAO, Rome, Italy; Belhaven Press, London, 338 pp.

Allden, W.G., 1970. The Effect of Nutritional Deprivation of the Subsequent Productivity of Sheep and Cattle. Nutr. Abstr. Rev., 40: 1167-1184.

Altona, R.E., 1966. Urea and Biuret as Protein Supplements for Range Cattle and Sheep in Africa. Outlook on Agriculture, 5: 22-27.

Andreae, B., 1980, The Economics of Tropical Agriculture. English Edition Edited by Kestner, J.M.A., Commonwealth Agricultural Bureaux, Oxford, UK. 178 pp.

Badurdeen, A.L., Ibrahim, M.N.M. and Schiere, J.B., 1994. Methods to Improve Utilization of Rice Straw I. Effects of Moistening, Sodium Chloride and Chopping on Intake and Digestibility. Asian-Aust. J. Anim. Sci., 7(2): 159-164.

Barton D., 1987. Draught Animal Power in Bangladesh. Ph.D. Thesis, School of development studies of the University of East Anglia.

Berckmoes, W.M.L., Jager, E.J. and Koné, Y., 1988. Intensification of Agriculture in South Mali. Wish or reality? Paper presented at the Farming Systems Research / Extension Symposium, University of Arkansas, 9-12 October 1988. Royal Tropical Institute, DRSPR. 27 pp.

Bieleman, J., 1987. Boeren op het Drentse zand 1600-1910. Een nieuwe visie op de 'oude' landbouw. Ph.D. Thesis, Wageningen / Utrecht. 834 pp.

Binswanger, H., 1986. Agricultural Mechanization; A Comparative Historical Perspective. Res. Observer, 1(1): 27-56.

Bosman, H.G. and Moll, H.A.J., 1995. Appraisal of Small Ruminant Production Systems; 3. Benefits of Livestock and Missing Markets: the Case of Goat Keeping in South-Western Nigeria (in print).

Budelman, A. and Van Der Pol, F., 1992. Farming System Research and the Quest for a Sustainable Agriculture. Agroforestry Systems, 19: 187-206.

Byerlee, D., Iqbal, M. and Fischer, K.S., 1989. Quantifying and Valuing the Joint Production of Grain and Fodder from Maize Fields: Evidence from Northern Pakistan. Expl. Agric., 25: 435-445.

Capper, B.S., 1990. The Role of Food Legume Straw and Stubble in Feeding Livestock. p.151-162. In: A.E. Osman et al. (eds.), The Role of Legumes in the Farming System of the Mediterranean Areas. ICARDA, The Netherlands.

Chayanov, A.V., 1926. The Theory of Peasant Economy. Reprinted and edited by D. Thorner et al., 1966. The American economic association translation series, Homewood, Illinois. 317 pp.

Clason, A.T., 1977. Jacht en Veeteelt. Unieboek, Haarlem.

Conway, G.R., 1985. Agroecosystem Analysis. Agric. Admin., 20: 31-55.

Conway, G.R. and Barbier, E.B., 1990. After the Green Revolution, Sustainable Agriculture for Development, Earthscan Publications, London. 205 pp.

Cox, G.W. and Atkins, M.D., 1974. Agricultural Ecology, an Analysis of World Food Production Systems. W.H. Freeman and Company, San Francisco. 721 pp.

Crotty, R., 1980. Cattle, Economics and Development. Commonwealth Agricultural Bureaux, Farnham Royal, Slough Sl2 3BN, 253 pp.

De Boer, J.A. and Welsch, D.E., 1977. Constraints on Cattle and Buffalo Production in a North Eastern Thai Village. p. 115-141. In: R.D. Stevens (ed.), Tradition and Dynamics in Small Farm Agriculture. Iowa state Unit Press.

De Lasson, A., 1981. Socio-Economic Aspects of Livestock Keeping in Noakhali District, Bangladesh. p. 456-495. In: M.G. Jackson, F. Dolberg, C.H. Davis, M. Haque and M. Saadullah (eds.), Maximum Livestock Production from Minimum Land. Bangladesh Agricultural University, Mymensingh, Bangladesh.

De Wit, J., Dhaka, J.P. and Subba Rao, A., 1993. Relevance of Breeding and Management for More or Better Straw in Different Farming Systems. p. 404-414. In: Kiran Singh and J.B. Schiere (eds.), Feeding of Ruminants on Fibrous Crop Residues: Aspects of Treatment, Feeding, Nutrient Evaluation, Research and Extension. Proc. of a Workshop, 4-8 february 1991, NDRI-Karnal. ICAR, Krishi Bhavan, New Delhi. Dept. Tropical Animal Production, Wageningen Agricultural University, Wageningen, 486 pp.

Durning, A.B. and Brough, H.B., 1991. Taking Stock: Animal Farming and the Environment. World Watch Paper 103.

George, P.S., Nair, K.N., Strebel, B., Unnithan, N.R. and Wälty, S., 1989. Policy Options for Cattle Development in Kerala. Submitted to the Kerala Livestock Development Board, Trivadrum, India and Intercooperation / Swiss Development Cooperation, Bern Switserland. 84 pp.

Harris, M., 1965. The Myth of the Sacred Cow. p. 217-218. In: A. Leeds and A.P. Vajda (eds.), Man, Culture and Animals, The Role of Animals in Human Ecological Adjustments. Publ. 78, AAAS, Washington, DC.

Hartley, B.S., F.R.S., Broda, P.M.A. and Senior, P.J. (eds.), 1987. Technology in the 1990's: Utilization of Lignocellulosic Wastes. Philos. Trans. R. Soc. London, 321: 403-568.

Jackson, M.G., 1983. A Strategy for Improving the Productivity of Livestock in the Hills of Uttar Pradesh. p. 130-154. In: Proc. of the Seminar: 'Regeneration of the Himalayas: Concepts and Strategies', 24-25 October 1983, Nainital, UP, India.

Jahnke, H.E., 1982. Livestock Production Systems and Livestock Development in Tropical Africa. Kieler Wissenschaftsverlag Vauk. 253 pp.

Janssen, H.G.P., Walker, T.S. and Barker, R., 1990. Adoption Ceilings and Modern Coarse Cereal Cultivars in India. American J. Agric. Econ., 653-663.

Jodha, N.S., 1986. Common Property Resources and Rural Poor in Dry Regions of India. Economic and Political Weekly; XXI, No.27, July 5: 1169-1181.

Joshi, A.L., Doyle, P.T. and Oosting, S.J., (eds.), 1994. Variation in the Quantity and Quality of Fibrous Crop Residues. Proc. Nat. Seminar, BAIF Development Research Foundation, Pune (Maharashtra-India), 8-9 February 1994. ICAR, New Delhi, India; Wageningen Agricultural University, Wageningen, 174 pp.

Kaasschieter, G.A., De Jong, R., Schiere, J.B. and Zwart, D., 1992. Towards a Sustainable Livestock Production in Developing Countries and the Importance of Animal Health Strategy Therein. The Veterinary Quarterly, 14 (2): 66-75.

Kang, B.T., Reynolds, L. and Atta-Krah, A.N., 1990. Alley Farming. Advances in Agronomy, vol. 43, 43 pp.

Kelley, T.G., Parthasarathy Rao, P. and Walker, T.S., 1991. The Relative Value of Cereal Straw Fodder in the Semi-Arid Tropics of India: Implications for Cereal Breeding Programs at ICRISAT. Resource Management Program, Economics Group, Progress Rep. 105. ICRISAT, Patancheru PO, Andhra Pradesh 502 324, India.

Kelley, J., 1992. Upgrading of Waste Cereal Straws. Outlook on Agriculture, 21(2): 105-108.

Ketelaars, J.J.M.H. and Tolkamp, B.J., 1992. Toward a New Theory of Feed Intake Regulation in Ruminants. 1. Causes of Differences in Voluntary Feed Intake: Critique of Current Views. Livest. Prod. Sci., 30: 269-296.

Kidane, H., 1984. The Economics of Farming Systems in the Different Ecological Zones of Embu District (Kenya), with Special Reference to Dairy Production. Ph.D. Thesis, Hannover University, Hannover, 227 pp.

Kiran Singh and Schiere, J.B. (eds.), 1993. Feeding of Ruminants on Fibrous Crop Residues: Aspects of Treatment, Feeding, Nutrient Evaluation, Research and Extension. Proc. of an Int. Workshop, 4-8 February 1991, National Dairy Research Institute, Karnal (Haryana-India), Indian Council for Agricultural Research, Krishi Bhavan, New Delhi, India; Dept. Tropical Animal Production, Wageningen Agricultural University, Wageningen. 486 pp.

Kiran Singh, and Schiere, J.B. (eds.), 1995. Handbook for Straw Feeding Systems; Principles and Applications with Emphasis on Indian Livestock Production. Indian Council for Agricultural Research, Krishi Bhavan, New Delhi, India and Department Tropical Animal Production, Agric. Univ. Wageningen. 428 pp.

Lal, R., 1988. Effect of Macrofauna on Soil Properties in Tropical Ecosystems. Agriculture, Ecosystems and Environment, 24: 101-116.

Lagemann, J., 1977. Traditional African Farming Systems in Eastern Nigeria; An analysis of Reaction to Increasing Population Pressure. Weltforum Verlag Munchen, (Afrika-Studien: No. 98).

Lamers J. and Feil, P., 1993. The Many Uses of Millet Residues. ILEIA Newsletter 9(3).

Lord Ernle, 1961. English Farming, Past and Present, 6th edn., Heinemann, London / Melbourne / Toronto, Frank Cass & Co, London. 559 pp.

McCourt, D., 1955. Infield and Outfield in Ireland. The Ec. Hist. Rev., Sec. Ser. 7: 369-276.

Nitis, I.M., Lana, K., Suarna, M. Sukanten, W. and Putra, S., 1991. Legume Trees as Alternative Feed Resources for Small Ruminants Integrated with Tree Crops. p. 24-33. In: L.C. Iniguez and M.D. Sanchez (eds.), Integrated Tree Cropping and Small Ruminant Production Systems. Proc. of a Workshop on Research Methodologies, 9-14 September 1990, Medan, North Sumatra. Gaya Technik, Bogor, Indonesia. 329 pp.

Nordblom, T.L., 1983. Livestock-Crop Interaction. The Decision to Harvest or to Graze Mature Grain Crops. Disc. Pap. No.10. International Centre for Agricultural Research in the Dry Areas (ICARDA), Farming Systems Program ICARDA, Aleppo, Syria. 21 pp.

Odum, H.T., 1971. Environment, Power and Society. Wiley Interscience, New York. 330 pp.

Overton, M., 1991. The Determinants of Crop Yields in Early Modern England, In: B.M.S. Campbell and M. Overton (ed.), Land, Labour and Livestock: Historical Studies in European Agricultural Productivity, Manchesta University Press, Manchester / New York.

Palthe, J.G.L., 1989. Upland Farming on Java, Indonesia: a Socio-Economic Study of Upland Agriculture and Subsistence Under Population Pressure. Netherlands Geographical Studies 97, Amsterdam / Utrecht.

Panayotou, T., and Tokrisna, R., 1982. Micro-Economics of Rural Livestock: The Case of Buffalo and Cattle in Thailand. p.65-74. In: J.C. Fine and R.G. Lattimore (eds.), Livestock in Asia: Issues and Policies. IDRC-202e.

Patil, B.R., Rangnekar, D.V. and Schiere J.B., 1993. Modelling of Crop-Livestock Integration: Effect of Choice of Animals on Cropping Pattern. p. 336-343. In: Kiran Singh and J.B. Schiere (eds.), Feeding of Ruminants on Fibrous Crop Residues: Aspects of Treatment, Feeding, Nutrient Evaluation, Research and Extension. Proc. of a Workshop, 4-8 february 1991, NDRI-Karnal. ICAR, Krishi Bhavan, New Delhi, India; Dept. Tropical Animal Production, Wageningen Agricultural University, Wageningen, 486 pp.

Pingali, P., Bigot, Y. and Binswanger, H.P., 1987. Agricultural Mechanization and the Evolution of Farming Systems in Sub-Saharan Africa. Published for the World Bank, The Johns Hopkins University Press, Baltimore and London, 216 pp.

Powell, J.M., 1985. Yields of Sorghum and Millet and Stover Consumption by Livestock in the Subhumid Zone of Nigeria. Trop. Agric. (Trinidad), 62(1): 77-81.

Preston, T.R. and Leng, R.A., 1987. Matching Ruminant Production Systems with Available Resources in the Tropics and Sub-Tropics. Penambul Books, Armidale, N.S.W.

Reed, J.D., Capper, B.S. and Neate, P.J.H. (eds.), 1988. Plant Breeding and the Nutritive Value of Crop Residues. Proc. of a Workshop, 7-10 December 1987, ILCA, Addis Ababa, Ethiopia, 334 pp.

Rifkin, J., 1992. Beyond Beef: the Rise and Fall of the Cattle Culture. A Dutton book, Penguin, New York, 353 pp.

Ruthenberg, H., 1980. Farming Systems in the Tropics. 3rd edn., Clarendon Press, Oxford.

Schiere, J.B., 1992. Veeteelt en Bodemvruchtbaarheid, Concentreren of Conserveren? In: Verslag van de vierenzestigste tropische landbouwdag; Nutriëntenstromen en duurzaam landgebruik in de tropen. 11 november 1992, Nili-KGvL, Wageningen; KIT, Amsterdam.

Schütz, H., Seiler, W. and Rennenberg, H., 1990. Soil and Land Use Related Sources and Sinks of Methane (CH_4) in the Context of the Global Methane Budget. In: A.F. Bouwman (ed.), Soils and the Greenhouse Effect, Present Status and Future Trends Concerning the Effect of Soils and their Cover on the Fluxes of Greenhousse Gases, the Surface Energy Balance and the Water Balance. Proc. Int. Conf. Soils and the Greenhouse Effect, 14-18 August 1989, Wageningen. International Soil Reference and Information Centre (ISRIC) and the Netherlands' Ministry of Housing, Physical Planning and Environment (VROM), 573 pp.

Shanmugartnam, N., Vedeld, T., Mossige, A. and Bovin, M., 1992. Resource Management and Pastoral Institution Building in the West African Sahel. World Bank Discussion Pap. No. 175. 77 pp.

Singh, R.B. and Saha, R.C., 1995. Jute and its By-Products. p.413-418. In: Kiran Singh, and Schiere, J.B. (eds.), 1995. Handbook for Straw feeding Systems; Principles and Applications with Emphasis on Indian Livestock Production. Indian Council for Agricultural Research, Krishi Bhavan, New Delhi, India and Department Tropical Animal Production, Agric. Univ. Wageningen. 418 pp.

Slicher Van Bath, B. H., 1963. The Agrarian History of Western Europe, A.D. 500-1850. Translated by O. Ordish. Arnold, London. 364 pp.

Spedding, C.R.W., 1988. An Introduction to Agriculture Systems, 2nd edn., Elsevier Applied Science.

Staniforth, A.R., 1982. Straw for Fuel, Feed and Fertilizer. Farming Press, Ipswich, Suffolk, 153 pp.

Steinfeld, H., 1988. Livestock Development in Mixed Farming Systems; A Study of Smallholder Livestock Production Systems in Zimbabwe. Farming systems and Resource Economics in the Tropics Vol. 3, Wissenschaftsverlag Vauk Kiel, 244 pp.

Thole, N.S., Sharma, D.D., De Wit, J. and Singhal, K.K., 1993. Feeding Strategies During Natural Calamities. pp.344-357. In: Kiran Singh and Schiere, J.B. (eds.) Feeding of ruminants on fibrous crop residues. Aspects of treatment, feeding, nutrient evaluation, research and extension. Proc. of a workshop, 4-8 february 1991, NDRI-Karnal. ICAR, Krishi Bhavan, New Delhi, India; Dept. Trop. Anim. Prod., Agricultural University, Wageningen. 486 pp.

Udo, H.M.J., Hermans C. and Dawood, F., 1990. Comparison of Two Cattle Production Systems in Pabna District Bangladesh. Trop. Anim. Health Prod., 22: 247-259.

Vaidyanathan, A., 1988. Bovine Economy in India. Centre for Development studies, Monograph series, Trivandrum.

Van Der Graaf, S., 1985. La lutte: a Study of the Consequences of Permanent Housing of the Cattle of the Mossi (Dutch). M.Sc. Thesis. Dept. Tropical Animal Production, Wageningen Agricultural University, Wageningen.

Wahed, R.A., Owen, E., Naare, M. and Hosking, B.J., 1990. Feeding Straw to Small Ruminants. Effect of Amount Offered on Intake and Selection of Barley Straw by Goats and Sheep. Anim. Prod., 51: 283-289.

Westgaard, P. and Sundstøl, F., 1986. History of Straw Treatment in Europe. p. 155-163. In: M.N.M. Ibrahim and J.B. Schiere (eds.), Rice Straw and Related Feeds in Ruminant Rations. Proc. Int. Workshop, Kandy, Sri Lanka, SUP Publication No.2. 407 pp.

Willerding, U., 1980. Anbaufrüchte de Eisenzeit und des frühen Mittelalters, ihre Anbau Formen, Standörtsverhältenisse und Erntemethoden, p. 126-196. In: H. Bech, D. Denecke and H. Jankuhn (eds.), Untersuchungen zur eizenzeitlichen und frühmittelalterlichen Flur in Mitteleuropa und ihrer Nutzung: Bericht über die Kolloquien der Kommission für die Alterumskunde Mittel- und Nordeuropas in den Jahren 1975 und 1976. Van Der Hoek & Ruprecht, Goettingen. 2 Tl.

Winrock, 1978. The Role of Ruminants in Support of Man. Winrock International Livestock Research and Training Centre, Morrilton, Arkansas. 136 pp.

Zemmelink G., 1986. Ruminant Production and Availability of Feed. p. 37-49. In: M.N.M. Ibrahim and J.B. Schiere (eds.), Rice Straw and Related Feeds in Ruminant Rations. Proc. Int. Workshop, 24-28 March 1986, Kandy, Sri Lanka. The Straw Utilization Project, Publication No.2, 408 pp.

Section 3

COMPONENT RESEARCH:
TREATMENT AND / OR SUPPLEMENTATION
OF STRAW BASED RATIONS

3.1.   Overcoming the nutritional limitations of rice straw for ruminants: response of growing sahiwal and local cross heifers to urea upgraded and urea supplemented straw

3.2.   Response of growing cattle given urea-treated and untreated rice straw, to supplementation with rice bran and lickblocks containing urea and molasses

3.3.   Overcoming the nutritional limitations of rice straw for ruminants: urea ammonia upgrading of straw and supplementation with rice bran and coconut cake for growing bulls

3.4.   Response of growing bulls to diets containing untreated or urea-treated rice straw with rice bran supplementation

The artificial preparation of gold is impossible:
> *... on peut conclure que les Chimistes qui travaille à en faire, doivent perdre inutilement leurs peines, & que ceux qui se vantent d'avoir ce secret, sont des charlatans & des fourbes, qui ne cherchent qu'à attraper des personnes credules.*
from Hartsoeker, p. 453. In: Partington, J.R., 1961 A History of Chemistry, volume two. Macmillan St. Martin's Press, London. 795 pp.

"One of the most highly developed skills in contemporary Western civilization is dissection: the split-up of problems into their smallest possible components. We are good at it. So good, we often forget to put the pieces back together again.
The skill is perhaps most finely honed in science. There we not only routinely break problems down into bite-sized chunks and mini-chunks, we then very often isolate each one from its environment by means of a useful trick. We say ceteris paribus - all other things being equal. In this way we can ignore the complex interactions between our problem and the rest of the universe."
> Alvin Toffler in the foreword (page XI) of Ilya Prigogine and Isabelle Stengers, 1985, "Order out of Chaos; Man's New Dialogue with Nature". Flamingo, London.

OVERCOMING THE NUTRITIONAL LIMITATIONS OF RICE STRAW FOR RUMINANTS:

RESPONSE OF GROWING SAHIWAL AND LOCAL CROSS HEIFERS TO UREA UPGRADED AND UREA SUPPLEMENTED STRAW[1]

J.B. Schiere and J. Wieringa

SUMMARY

Thirty-six heifers of three different breeds were fed rice straw, either upgraded with 4% urea, or supplemented with 2% urea, sprayed on the straw just prior to feeding. The effects on liveweight gain and dry matter intake were measured for pure Sahiwal heifers, Sahiwal x local crosses and Jersey x local crosses. Urea upgraded straw gave better growth than urea supplemented straw, average 217 g/day/animal *versus* 71 g/day/animal. This was associated with a higher intake of upgraded straw, compared to supplemented straw (2.4 *versus* 1.8 kg/100 kg BW). No overall breed effect on growth was found (P > 0.05).

---

[1] Published in the Asian-Australian Journal of Animal Science, 1988, 1(4): 209-212.

INTRODUCTION

Rice straw is a major feed resource for ruminants in many tropical countries, especially during the dry season. Despite frequently occurring shortages of roughage in Sri Lanka the straw is often burned in the field for disposal. Straw contains too little digestible energy and protein to sustain even maintenance of animals (O'Donovan, 1983).

There are two ways to overcome this deficiency of nutrients. The first method is to upgrade the straw through treatment with urea which is converted into ammonia (Perdok *et al.*, 1982; Ibrahim, 1983). Alternatively, the deficient nutrients may be provided through supplements, such as concentrates, urea or immature green forages (Creek *et al.*, 1984, Ghebrehiwet *et al.*, 1988).

In this experiment, the effect of urea upgrading *versus* supplementation with urea (sprayed on straw) on liveweight gain and intake was studied for heifers of three different breeds.

MATERIALS AND METHODS

Treatments

Rice straw supplemented with 2% urea, which was sprayed on the straw just prior to feeding, was compared with rice straw upgraded with 4% urea under airtight conditions. Each ration was fed to six growing heifers of three different breeds, i.e. Sahiwal, Sahiwal x Local cross and Jersey x Local cross. This resulted in six groups of six animals each, 12 of each breed.

The 12 pure Sahiwals varied in age from 11 to 29 months and in weight from 90 to 200 kg liveweight. The 12 Sahiwal crosses and 12 Jersey crosses were about one year old and varied in weight from 60 to 90 kg. The three groups came from different farms within the Coconut Triangle in Sri Lanka. The 12 animals of each breed were allotted homogeneously in regard to body weight to two ration groups over two stables. The animals were housed back to back in open two-row sheds.

Feeds and feeding

The basal feed was rice straw, obtained from village farmers and fed unchopped. The straw was either supplemented(sprayed) or upgraded with urea.

The urea supplement was given to the animals by adding a 2% solution of urea to the straw just prior to feeding without allowing time for reactions between urea and the straw. After putting straw in the feed trough, it was sprayed with 100 l urea solution / 100 kg airdry straw, resulting in 2.0 kg urea / 100 kg airdry straw.

The upgraded straw was produced by addition of 4 kg urea in 100 l water to 100 kg airdry straw allowed to react for 9-11 days in a concrete pit sealed with polythene (Schiere *et al.*, 1988). The straw was mixed with the urea solution in the pit itself using watering cans.

After nine days, the upgraded straw was fed over the next three days. On the 12th day, a new lot of upgraded straw was started that had been made on the fourth day.

Straw was fed *ad libitum*, keeping the feed troughs full day and night and removing refusals every morning. In addition to the experimental rations, all animals were fed 1 kg of fresh grass (cut in the field irrespective of maturity). The grass (unchopped) was offered on top of the straw in the feed troughs. Also given was a daily supplement of 0.5 kg local rice bran, and 20 g sodium sulphate, 10 g di-calcium phosphate and 30 g mineral mix. The animals had free access to drinking water.

Measurements

The experiment lasted for 11 weeks, consisting of an adaptation period of three weeks and a measurement period of eight weeks. Body weights were determined before feeding at weekly intervals using a cattle scale. Dry matter intake (DMI) of the animals was estimated for each group of three animals, by weighing feed offered and feed refused during five days, so for each ration group there were two observations. Samples of feed offered and refused were taken and analyzed for dry matter at Peradeniya University. Due to distance and logistical problems, the analyses were not carried out immediately, so dry matter contents may have been overestimated.

Statistical analysis

Liveweight gain and dry matter intake were analyzed using three-way analysis of variance (Snedecor and Cochran, 1980), with type of straw (upgraded, supplemented), breed (Sahiwal, Sahiwal x Local and Jersey x Local) and stable (1,2) as main effects. Mean rate of liveweight gain (LWG) was calculated by means of linear regression analysis (Snedecor and Cochran, 1980).

RESULTS AND DISCUSSION

Liveweight gain and dry matter intake of animals on urea upgraded straw were significantly (P < 0.01) higher than on urea supplemented straw(see table 1). On an average, the animals on upgraded straw grew 217 g.d^{-1} *versus* 71 g.d^{-1} for the animals on supplemented straw. The straw intakes were 2.4 and 1.8 kg/100 kg BW, respectively. The stable effect and the effect of initial weight (as a covariable) were not significant (P > 0.05).

The difference in DMI and LWG between urea upgraded and supplemented straw as found in this experiment agrees with an experiment of A. De Rond and colleagues (unpublished data), who showed that results are dependent on the level of urea used. They compared untreated straw with urea upgraded and urea supplemented straw, using 2%, 4% and 6% urea solutions. At the two highest levels (4% and 6%) a significant difference in dry matter intake resulted between upgraded and supplemented straw, while intakes were almost equal at a level of 2% urea. As optimum treatment levels they found 2% urea when supplied as a supplement and 4% urea when upgrading straw. The nutritional superiority of upgraded straw (4% urea) over supplemented straw (2% urea) was also shown by Van Der Hoek *et al.*(1989) who found a higher milk and butterfat production, as well as less liveweight loss of lactating Surti buffaloes, when fed the upgraded straw. The higher growth rate on upgraded straw was associated with a higher intake of urea upgraded straw compared with the urea supplemented straw, as also shown by Jaiswal *et al.* (1983) Karunaratne and Jayasuriya (1984) and Perdok *et al.* (1984). It might also be caused by the fact that upgraded straw has a higher dry matter digestibility (Hossain and Rahman, 1981; Karunaratne and Jayasuriya, 1984), and a crude protein content exceeding 7% (Schiere and Ibrahim, 1985), compared with 4% in untreated straw (Doyle *et al.*, 1986) and an intermediate content in straw supplemented with 2% urea. Jayasuriya and Perera (1983) even found crude protein contents of upgraded straw as high as 11-13% in the dry matter, of samples that had not been ovendried before crude protein determination. By ovendrying part of the urea is lost in the form of gaseous ammonia, which underestimates the actual crude protein content of fresh upgraded straw.

Table 1.   Effect of breed and type of straw on liveweight gain and intake of heifers receiving urea upgraded rice straw or rice straw, supplemented with 2% urea

	Sahiwal		Sahiwal x Local		Jersey x Local	
	Upgraded	Suppl. 2%	Upgraded	Suppl. 2%	Upgraded	Suppl. 2%
Liveweight gain (g/day)	282[a]	105b[c]	185[b]	70[c]	183[b]	39[c]
Dry matter intake (kg/100kg BW)[2]						
Straw	2.33[a]	1.89[b]	2.49[a]	1.83[b]	2.59[a]	1.70[b]
Grass	0.13	0.14	0.25	0.26	0.22	0.25
Rice Bran	0.29	0.31	0.54	0.56	0.48	0.54
Total	2.75	2.34	3.27	2.65	3.30	2.49

[1]a,b,c   Values with the same superscripts are not significantly different (P > 0.05).
[2]   Estimated.

Doyle *et al.* (1986) reported an experiment with sheep, which attempted to partition the benefit of upgrading into that caused by the higher nitrogen content and that by the chemical reaction of ammonia with cell wall components. Urea supplementation of untreated straw at a rate of 1.2% of dry matter intake increased intake of digestible organic matter from 270 to 430 g.d^{-1}, while upgrading with urea resulted in an intake of 480 gd^{-1}. The intakes of nitrogen on both rations were equal (12 g.d^{-1}). They concluded, that appropriate supplementation with urea, under ideal conditions, accounted for 75% of the increase in nutritive value of straw by the treatment reaction *per se* (Doyle *et al.*, 1986).

No overall effects of breed on liveweight gain and dry matter intake were found (P > 0.05). Breed straw type interactions were not found either (P > 0.05), although liveweight gain of pure Sahiwals on upgraded straw was higher (P < 0.05) than liveweight gain of both crosses on upgraded straw (table 1). The better growth of Sahiwals may be due to a different growth stage or life history of the crosses. Confounding of breed effect and life history/origin of the animals is possible, since the three groups came from different farms.

Whether treatment is economically justified depends on beef prices and cost of inputs. Also important are the hidden benefits of urea upgraded straw, such as better health and probably younger age of first calving. The economical evaluation of urea upgraded straw as a cattle feed has been elaborated by Nell *et al.* (1986) and Schiere *et al.* (1988).

ACKNOWLEDGEMENTS

This paper reports an experiment conducted in 1983 at NLDB Galpokuna farm, Sri Lanka by the Straw Utilization Project, supported by the National Livestock Development Board, University of Peradeniya, Mahaweli Authority of Sri Lanka and the Directorate General of International Cooperation of the Netherlands. Thanks are due to H.B. Perdok (Straw Treatment Project), to M.C.N. Jayasuriya (University of Peradeniya), to the farm manager, the farm assistants Padmanadan and Edwin and the labourers of NLDB Galpokuna farm, and to V.J.H. Sewalt, who did the editing of this paper at the Department of Tropical Animal Production of the Agricultural University in Wageningen, the Netherlands.

REFERENCES

Creek, M.J., T.J. Barker and W.A. Hargus. 1984. The development of a new technology in an ancient land. World Animal Review 51: 12-20.

Doyle, P.T., C. Devendra and G.R. Pearce. 1986. Rice straw as a feed for ruminants. International Development Program of Australian Universities and Colleges (IDP), Canberra, Australia, p. 117

Ghebrehiwet, T., M.N.M. Ibrahim and J.B. Schiere. 1988. Response of growing bulls to diets containing untreated or urea-treated rice straw with rice bran supplementation. Biological Wastes 25: 269-280.

Hoek, Van Der, R., G.S. Muttetuwegama and J.B. Schiere. 1989. Overcoming the nutritional limitations of rice straw. 1. Urea ammonia treatment and supplementation of straw for lactating Surti buffaloes. Asian-- Australian Journal of Animal Sciences 1: 201-208.

Hossain, S.A. and M.S. Rahman. 1981. Comparative feeding value of urea treated and untreated paddy straw. In: Preston, T.R., F. Dolberg, C.H. Davis, M. Haque and M. Saadullah (eds.), Maximum Livestock Production from Minimum Land, Proceedings of a seminar held in Bangladesh, p. 225-233.

Ibrahim, M.N.M. 1983. Physical, chemical, physico-chemical and biological treatments of crop residues. In: Pearce, G. (ed.), The utilization of fibrous agricultural residues. Proceedings of a seminar held in Los Banos, Philippines, 18-23 May 1981. Australian Government. Publishing Service, Canberra, p. 53-68.

Jaiswal, R.S., M.L. Verma and I.S. Agarwal, 1983. Effect of urea and protein supplements added to untreated and ammonia-treated rice straw on digestibility, intake and growth of crossbred heifers. In: Davis, C.H., T.R. Preston, M. Haque and M. Saadullah (eds.), maximum livestock production from minimum land. Proceedings of the fourth seminar, held in Bangladesh 1983, p. 26-31.

Jayasuriya, M.C.N, and H.G.D. Perera. 1983. Urea-ammonia treatment of rice straw to improve its nutritive value for ruminants. Agricultural Wastes 4: 143-150.

Karunaratne, M. and M.C.N. Jayasuriya. 1984. Ruminal digestion of diets containing urea-supplemented straw or straw treated with sodium hydroxide or ammonia released from urea. In: Doyle, P.T. (ed.), The utilization of fibrous agricultural residues as animal feeds. Proceedings of the third annual workshcp of the AAFARR Network. School of Agriculture and Forestry, University of Melbourne, Parkville, Victoria, p. 166-171.

Nell, A.J., J.B. Schiere and M.N.M. Ibrahim. 1986. Economic evaluation of urea-ammonia treated straw as cattle feed. In: Ibrahim, M.N.M. and J.B. Schiere (eds.), Rice straw and related feeds in ruminant rations. Proceedings of an international workshop held in Kandy, Sri Lanka, 24-28th March, 1986. Straw Utilization Project, Kandy, Sri Lanka, p. 164-170.

O'Donovan, P.B. 1983. Untreated straw as a livestock feed. Nutrition Abstracts and Reviews - Series B 53: 441-455.

Perdok, H.B., G.S. Muttetuwegama, G.A. Kaasschieter, H.M. Boon, N.M. Van Wageningen, V. Arumugam M.G.F.A. Linders, and M.C.N. Jayasuriya. 1984. Production responses of lactating or growing ruminants fed urea-ammonia treated paddy straw with or without supplements. In: Doyle, P.T. (ed.), The utilization of fibrous agricultural residues for animal feeds. Proceedings of the third annual workshop of the AAFARR Network. School of Agriculture and Forestry, University of Melbourne, Parkville, Victoria, p. 213-230.

Preston, T.R. and R.A. Leng. 1984. Supplementation of diets based on fibrous residues and by-products. In: Sundstøl, R.F. and E. Owen (eds.), Straw and other fibrous by-products as feed. Developments in Animal and Veterinary Sciences 14. Elsevier, Amsterdam, p. 373-413.

Schiere, J.B. and M.N.M. Ibrahim. 1985. Recent research and extension on rice straw feeding in Sri Lanka: A review. In: Doyle, P.T. (ed.), The utilization of fibrous agricultural residues as animal feeds. Proceedings of the fourth annual workshop of the, AAFARR Network held in Khon Kaen, Thailand, 10-14 April, 1984. IDP, Canberra, p. 116-126.

Schiere, J.B., A.J. Nell and M.N.M. Ibrahim. 1988. Feeding of urea-ammonia treated rice straw. World Animal Review 65: 31-42.

Snedecor, G.W. and W.G. Cochran, 1980. Statistical methods (7th Edition). Iowa State University Press, Ames, Iowa, USA.

Chapter 3.2

RESPONSE OF GROWING CATTLE GIVEN UREA-TREATED AND UNTREATED RICE STRAW TO SUPPLEMENTATION WITH RICE BRAN AND LICKBLOCKS CONTAINING UREA AND MOLASSES[1]

J.B. Schiere, M.N.M. Ibrahim, V.J.H. Sewalt and G. Zemmelink

SUMMARY

In an experiment with 48 growing Sahiwals (both bulls and heifers) the effect of access to a urea-molasses lickblock on straw diets was studied. The animals were given rice straw of unknown variety either untreated (US), supplemented with rice bran and concentrates (USRB) or treated with urea-ammonia (TS). Within each diet, animals were given or not given access to lickblocks containing urea, molasses, minerals and cottonseed meal. Individual dry matter intake (DMI) was measured daily during two periods of 8 days. Dry matter digestibility (DMD) was determined by using acid-insoluble ash (AIA) as an indigestible marker. Results were analyzed statistically with ration (US, USRB and TS) and lickblock (-, +) as main effects. The effect of lickblock supplementation on straw DMI, total DMI and DMD was not significant (P > 0.05). Straw intake was significantly higher (P < 0.001) for TS (101.2 g kg $BW^{-0.75}$ per day) than for US (79.5 g kg $BW^{-0.75}$ per day). There was no clear substitution effect of RB intake on straw intake (P > 0.05), so that total DMI (85.8 g kg $BW^{-0.75}$ per day for US) was increased by RB supplementation to 97.3 g kg $BW^{-0.75}$ per day (P < 0.01). Total intake was highest for the TS groups (106.1 g kg $BW^{-0.75}$ per day). DMD of TS (55.5%) significantly (P < 0.01) differed from DMD of US (46.3%) and DMD of USRB (48.3%). No effect of lickblock on LWG was found (P > 0.05). Urea treatment significantly (P < 0.05) increased LWG g per day from -111 g per day (average US diets) to +83 g per day (average TS diets). The animals supplemented with rice bran showed an intermediate but not significantly different (P > 0.05) growth.

---

[1] Published in Animal Feed Science and Technology, 26(1989): 179-189.

INTRODUCTION

Straw and other crop residues are widely used as cattle feed, but they are inadequate as a sole source of nutrients. Straw quality could be improved by treatment with urea or other chemicals, resulting in higher digestibility and a higher intake (Saadullah *et al.*, 1981; Chesson and Ørskov, 1984; Ghebrehiwet *et al.*, 1988).

Supplementation with specific nutrients may overcome dietary inadequacies. Low levels of supplementation may have a beneficial effect on rumen fermentation, enhancing both the rate and extent of fermentation, and often increasing intake of straw (Preston and Leng, 1984). Many authors have reported positive effects of nitrogen, phosphorus, sulphur and carbohydrate supplementation on rumen fermentation and intake (Campling *et al.*, 1962; Coombe and Tribe, 1962; Ernst *et al.*, 1975; Leng, 1984a). An efficient way of adding nitrogen is the use of urea, which can be sprayed on the straw directly, or made available in a mixture with other feeds, such as molasses. This mixture can be given in a liquid or solid form (urea-molasses lickblock) and also additives such as phosphorus, sulphur, etc., could be incorporated. For an efficient rumen fermentation it is essential that these supplements are continuously ingested. The use and manufacture of lickblocks has been described by a number of authors both recently (Kunju, 1984, 1986; Leng, 1984b; Sudana, 1985; APHCA, 1986; Manget Ram and Kunju, 1986; Sansoucy, 1986), and in previous decades (Ministry of Agriculture, 1957; Altona, 1966; Loosli and McDonald, 1968; Alexander, 1972).

While a small amount of supplements may stimulate rumen function and digestion and intake of straw, high levels of supplements may lead to a depression of rumen function and a lower intake of straw (substitution effect), depending on composition and proportion of both basal roughage and supplement.

A commercially available block was evaluated with three different basal rations, untreated straw (US), untreated straw with additional rice bran concentrates (USRB) and urea-treated straw (TS), and the effects of lickblock on intake, digestibility and liveweight gain were recorded.

MATERIALS AND METHODS

Treatments

Forty-eight animals were allocated to six treatment groups as follows: US: untreated straw only; US$^+$: US + lickblock; USRB: US + 1 kg mixed concentrates (RB), of which 80% rice bran; USRB$^+$: US + 1 kg RB + lickblock; TS: treated straw only; TS$^+$: TS + lickblock. Each treatment group consisted of eight animals. The animals were grouped according to body weight and previous growth rate.

Animals and Housing

The animals used were pure Sahiwals and Sahiwal crossbreds, both young bulls and heifers, 8-18 months of age, weighing 100-200 kg. They were eartagged and treated with anthelmintics before the experiment started. The animals were housed in a half-walled shed equipped with individual feed troughs. The sticky lickblocks were presented to the animals on small concrete platforms to prevent contamination of the blocks with straw. The six treatment groups were randomized through the shed to avoid confounding stable and treatment effects.

Feeds and feeding

The basal feeds were untreated straw (US) and urea-treated straw (TS). Straw was obtained from village farmers and was fed unchopped. The TS was treated for 7-14 days, using a 4% urea solution with a water:straw ratio of 1:1 (Schiere and Ibrahim, 1989). This level of urea is similar to that recommended in India, but the quantity of water was slightly higher (ICAR, 1985). Straw was fed *ad libitum* (>20% excess feed), keeping the feed troughs full day and night and removing refusals every morning.

The lickblocks were continually available to the animals. The ingredients used and chemical composition of the blocks are given in Table 1. Rice bran was obtained from a local mill and was fed separately in wooden boxes. Because of the difficulty in initial acceptance, the rice bran was mixed with a commercial dairy concentrate in the proportion 80/20, respectively. This mixture will hereafter be referred to as RB. All animals were fed 1 kg of fresh grass (*Pennisetum purpureum* and *Brachiaria brizantha*, dry matter about 15%) each morning. The grass (unchopped) was offered on top of the straw in the feed troughs. During the adaptation period all animals were fed 30 g mineral mixture, to avoid a compensatory intake effect of lickblocks on mineral-deficient animals. During the measurement period no additional minerals were offered. The animals had free access to drinking water.

Table 1.   Ingredients and chemical composition (in dry matter) of the lickblock

Ingredients	(%)	Chemical composition	(%)[1]
Dry matter (DM)	97 (94.2)	Ash	28 (28.9)
Molasses	45	AIA	2 (2.6)
Urea	15	Calcium	6
Mineral mixture	15	Phosphorus	2
Salt	8	Crude protein	56 (60.7)
Binders		Ether extract	0.5
Calcite powder	4	Crude fat	3
Bentonite	3		
Cottonseed meal	10		

[1] Values are given by the manufacturer; those in parentheses are from our own laboratory.

Measurements of feed intake and digestibility

The experiment lasted for 12 weeks, consisting of a 6-week adaptation period and a 6-week measurement period. Individual straw dry matter intake was measured during two periods, one of 14 days and one of 10 days, respectively. For both periods the intakes of the last 8 days were used to calculate individual intake. Straw refusals were collected at 07.00 h and weighed individually. Samples of offered and refused straw were dried at 70°C for 24 h to estimate dry matter content. It was not practical to measure lickblock intake daily. Thus lickblock intake per animal was determined as the total intake during 14 days (first period) or 10 days (second period) divided by the number of days. Samples were taken from other lickblocks (by crushing them) and oven dried at 70°C during 5 days to determine dry matter. At the end of each period the remaining blocks were weighed.

The RB was offered in two parts: 500 g in the morning, and 500 g in the afternoon. Refused RB was weighed separately for each individual animal. After weighing, refusals of the two treatment groups (USRB⁻ and USRB⁺) were mixed and subsampled to determine dry matter. Feed samples (straw, lickblock, RB and grass) were analyzed for total ash according to the standard procedures of the Association of Official Analytical Chemists (Williams, 1984). Organic matter intake (OMI) was calculated using the dry matter intake of the separate feed components and their respective organic matter contents. Faecal samples were not analyzed for organic matter. Dry matter digestibility was estimated by using acid-insoluble ash (AIA) as an indigestible marker. From each treatment group, three animals were selected on the basis of easiness of handling and low variability in feed intake. Faecal samples were collected by grab sampling at 10.00 h during 8 days in the first intake-measurement period. The samples were stored in bottles and kept in a freezer. After the collection period the samples were dried, ground and a subsample was analyzed for AIA (Van Keulen and Young, 1977). Similarly, feed samples (straw, lickblock, RB and grass) were analyzed for AIA.

Liveweight gain measurement

The original intention was to measure liveweight gain for at least 3 months. However, due to shortage of straw and problems encountered with the weighing bridge, the experimental period was reduced to 6.5 weeks. Therefore measurements of liveweight gain were restricted to weight changes after 45 days.

Statistical analysis

Intake, digestibility and liveweight gain were analyzed using two-way analysis of variance (Snedecor and Cochran, 1980) with ration (US, USRB, TS) and lickblock (-,+) as main effects. Lickblock intake was used as a covariate. The Bonferroni test (Neter and Wasserman, 1974) was used to test differences between treatment groups.

RESULTS

The effect of period on intake and interactions between period and the main effects (ration and lickblock) were not significant (P > 0.05). Therefore, the analysis was based on the mean values for both periods. A summary of the intake data is presented in Table 2.

Table 2.   Dry and organic matter intakes of bull calves and heifers given a basal diet of rice straw (US⁻), or rice straw supplemented with: urea-molasses lickblock (US⁺), 1 kg rice bran concentrates (USRB⁻), urea-molasses lickblock and 1 kg rice bran concentrates (USRB⁺), or given urea-treated rice straw (TS⁻), or urea-treated rice straw supplemented with a urea-molasses lickblock (TS⁺)[1]

Treatment	Dry matter intake (g $kg^{-0.75}$/day)				Organic matter intake (g $kg^{-0.75}$/day)
	Straw	Lick-block	Rice bran	Total	
US⁻	79.2[a]	-	-	82.9[a]	71.0[a]
	(4.4)	-	-	(4.5)	(3.9)
US⁺	79.8[a]	5.4[a]	-	88.7[a]	75.2[a]
	(6.9)	(1.9)	-	(8.4)	(7.0)
USRB⁻	75.1[a]	-	18.1[a]	96.8[b]	81.9[b]
	(4.1)	-	(6.4)	(9.1)	(7.5)
USRB⁺	74.7[a]	3.9[a]	15.6[a]	97.8[b]	82.2[b]
	(4.5)	(1.6)	(5.1)	(5.5)	(4.5)
TS⁻	101.5[b]	-	-	104.9[c]	89.9[c]
	(6.3)	-	-	(6.4)	(5.5)
TS⁺	100.8[b]	3.3[a]	-	107.3[c]	91.5[c]
	(3.8)	(1.9)	-	(3.3)	(2.8)

[1]   Values within parentheses are standard deviations (SD).
a,b,c: Values within the same column, followed by the same letters are not significantly different (P > 0.05).
[2]   Including grass supplement.

The interaction between lickblock and ration DMI, DMD and LWG was not significant (P > 0.05). Therefore, the final analysis was limited to testing the main effects only. The mean lickblock intake (g $kg^{-0.75}$ per day) for the group of animals receiving treated straw was 3.3 and that for animals receiving untreated straw without rice bran was 5.4. The differences between these two values was almost significant (P > 0.05). The intake of lickblock by individual animals varied from 1 to 8 g DM g $kg^{-0.75}$ per day, but the effect of varying intake of lickblock (covariable) on intake of straw was not significant (P > 0.05).

The dry matter intake of treated straw (101.2 g kg $^{-0.75}$) was 27% higher (P < 0.01) than the intake of untreated straw (79.5 g $kg^{-0.75}$). Supplementation with rice bran caused a slight, but not significant (P > 0.05) reduction in intake of straw (74.9 *vs.* 79.5 g $kg^{-0.75}$), representing a substitution rate of 27%. Straw intake of animals receiving lickblock (mean of all three treatment groups 85.1 g $kg^{-0.75}$ per day) was the same as for animals not receiving lickblock (85.3 g $kg^{-0.75}$ per day). In contrast with straw intake, total dry matter intake (g $kg^{-0.75}$ per day) increased from 85.8 to 97.3 as a result of supplementation with rice bran (P < 0.01). However, animals on treated straw consumed significantly (P < 0.01) more total dry matter (106.1) than animals receiving untreated straw and rice bran (P < 0.01). The mean total intake of animals receiving lickblock (97.9) was not significantly (P > 0.05) higher than that of animals not receiving lickblock (94.9). The results were similar for total intake of organic matter.

Dry matter digestibility (DMD) and liveweight gain (LWG) data are presented in Table 3. Assuming that the average DMD of three animals gives a fair estimate of the treatment group (eight animals), total intake of digestible dry matter (I_{DDM}) and the ratio nitrogen/digestible organic matter (N/DOM) were estimated (Table 3). Dry matter digestibility of treated straw (55.5%) was significantly (P < 0.01) higher than that of untreated straw (46-48%). Digestibility of total ration was not affected by the inclusion of rice bran or access to a lickblock (P > 0.05). Differences in intake of total digestible dry matter, due to lickblock supplementation, were small (47.3 vs. 49.8 g $kg^{-0.75}$ per day). Animals receiving treated straw consumed the most digestible dry matter, animals on untreated straw supplemented with RB were intermediate and animals receiving untreated straw without RB consumed the least digestible dry matter. These differences were calculated using average intake and digestibility per treatment group. The considerable effect of lickblock on N/DOM ratio was not reflected in increased intake. The N/DOM ratio should be around 0.032 g N g^{-1} DOM (ARC, 1980).

Table 3.　　　Dry matter digestibility (DMD), total intake of digestible dry matter (I_{DDM}), nitrogen/digestible organic matter (N/DOM) ratio of the diet and liveweight gain (LWG) of bull calves and heifers given a basal diet of rice straw (US⁻), or rice straw supplemented with: urea-molasses lickblock (US⁺), 1 kg rice bran concentrates (USRB⁻), urea-molasses lickblock plus 1 kg rice bran concentrates (USRB⁺), or given urea-treated rice straw (TS⁻), or urea-treated rice straw supplemented with urea-molasses lickblock (TS⁺)[1]

Treatment	DMD (%)	I_{DDM} (g $kg^{-0.75}$/day)	LWG (g/day)	N/DOM ratio
US⁻	47.3[a] (1.8)	39.2 (3.8)	-101[a] (64)	0.017
US⁺	45.4[a] (3.9)	40.3 (7.8)	-121[a] (64)	0.029
USRB⁻	46.3[a] (3.4)	44.8 (8.1)	+36[ab] (110)	0.021
USRB⁺	50.3[a] (1.4)	49.2 (4.3)	-17[ab] (112)	0.028
TS⁻	55.1[b] (1.8)	57.8 (5.6)	+76[b] (74)	0.037
TS⁺	55.8[b] (1.9)	59.9 (4.0)	+89[b] (78)	0.042

[1]　Values within parentheses indicate standard deviation (SD).
a,b: values within the same column followed by the same letter are not significantly different (P > 0.05).

Access to lickblock did not improve liveweight gain (P > 0.05). The animals on untreated straw lost weight (-111 g per day), the animals on untreated straw with RB maintained body weight (+10 g per day) and the animals on treated straw gained weight (+83 g per day). The liveweight changes are in agreement with the calculated I_{DDM} values. The intake of the animals on untreated straw, supplemented with RB (47 g $kg^{0.75}$ per day), was just enough for body maintenance.

DISCUSSION

Kunju (1986) reported lickblock intakes of 560 g per day (7.8 g $kg^{-0.75}$ per day) for animals of 300 kg, receiving rice straw as the basal ration and 530 g per day (6.5 g kg $^{-0.75}$ per day) for animals of 350 kg, receiving rice straw and 1 kg concentrates. Intake of the same commercially available lickblock was about 30-40% lower for animals on similar rations in our experiment. Kunju (1986) reported an increase in intake of straw from 4.4 to 5.7 kg per day, when he replaced 1 kg concentrates with 560 g lickblock. He also reported another trial, in which intake of straw marginally increased from 6.4 to 6.8 kg per day, when lickblock was added to a ration including 1 kg concentrates. These effects could not be clearly explained due to confounding of possible stimulation of straw intake by lickblock and substitution of straw by concentrates. In this experiment animals ate lickblock in addition to the same amount of straw, but intake of straw did not increase (P > 0.05). Although intake of straw did not significantly decrease as a result of supplementary RB, the mean intake of straw for the supplemented groups (74.9 g kg $^{-0.75}$ per day) was lower than that of the animals not receiving RB (79.5 g $kg^{-0.75}$ per day), representing a substitution rate of 0.27 g of straw g^{-1} rice bran. Lickblock did not substitute straw. Its failure to increase straw intake and digestibility may be due to the low level of consumption of lickblock, although even at the intakes measured the N content of the total ration was increased considerably. Although the blocks used in this experiment were 1 year old, the difference in results cannot be due to a change in chemical composition (Table 1). It may be due to the block changing its hardness when exposed, possibly affecting intake, but not chemical composition. High variation between animals may be caused by irregular intake of the block, which has been observed even under stall-feeding conditions (Manget Ram and Kunju, 1986).

Several workers have found increased intakes of the basal ration as a result of urea/molasses supplementation (Ernst *et al.*, 1975; Losada *et al.*, 1979; Sudana, 1985). Others, however, found no increased intakes of the basal ration (Chicco *et al.*, 1972; Church and Santos, 1981; Dixon, 1984; Neric *et al.*, 1985). Pearce (1973) found no increased hay intake, although liveweight gain increased. Effects on liveweight gain were more pronounced than effects on feed intake in the work of Kunju (1986) also. McLennan *et al.* (1981) showed that urea supplementation increased OMI by 14%, whereas additional molasses, sodium sulphate or both had no effect on intake. In that case, addition of readily available carbohydrates had little effect, so nitrogen probably was the primary deficiency.

Effects of lickblock supplementation on digestibility are doubtful. Also in other experiments, no positive effects of urea/molasses supplementation were found (Ernst *et al.*, 1975; Church and Santos, 1981). Soetanto *et al.* (1987) showed a positive effect of lickblocks on rate of degradation of dry matter and cell wall content. Whether the blocks are beneficial will depend on type of basal ration and level and type of supplement. Such conclusions were reached by Pearce (1973), who observed that the effects of lickblocks were more effective when the quality of the basal ration was poorer. At high levels of lickblock intake such as 2 kg per animal per day (O.A. El Khidir, personal communication, 1988), the lickblock becomes a supplement, rather than a stimulant for rumen function.

Due to the short period during which liveweight was measured, firm conclusions on liveweight gain cannot be drawn. Data indicate that liveweight gain was not affected by lickblock, while the differences between main rations appear clear. This agrees with Ghebrehiwet *et al.* (1988), who found a liveweight gain of -123 g per day for animals receiving untreated straw, +93 g per day for animals receiving treated straw and intermediate growth rates for animals on untreated straw supplemented with rice bran.

The variability in results implies the necessity for caution with regard to conclusions. Present data provide insufficient basis for the conclusion that expensive lickblocks would be beneficial. The inclusion of a small amount of green grass, providing not only some supplementary plant protein but also readily available carbohydrates, may be a cause for the smaller effect of lickblock and would in many instances be a more economic alternative for the farmer. Inclusion of 20% commercial rice bran (commonly available in Sri Lanka) did not increase the digestibility of the ration, but resulted in a somewhat higher total intake. The largest improvement in terms of both intake and digestibility was obtained when untreated rice straw was replaced by treated rice straw. When looking for possibilities of improving the nutrition of ruminants, this alternative deserves major attention.

ACKNOWLEDGEMENTS

This paper reports data from an experiment conducted by the Straw Utilization Project of Sri Lanka during September-December 1985. Thanks are especially due to C. De Saram and to R. Wijetunge of the Mahaweli Authority of Sri Lanka (Draught Animal Program), to the Directorate General of International Cooperation of The Netherlands for funding this research, to the manager of Niraviya Farm, to H. Blauw, P. Handunge, A.L. Badurdeen, and U.H.L. Chandra for doing analysis, and to B. Brouwer of the Department of Tropical Animal Production of the Agricultural University of Wageningen for assisting in the statistical analysis and giving other valuable comments.

REFERENCES

Alexander, G.I., 1972. Non-protein nitrogen supplements for grazing animals in Australia. World Anim. Rev., 11: 11-14

Altona, R.E., 1966. Urea and biuret as protein supplements for range cattle and sheep in Africa. Outlook Agric., 5: 22-27.

APHCA, 1986. Asian Livestock, Issue on Lickblocks. FAO Regional Office, Bangkok.

ARC, 1980. The nutrient requirements of ruminant livestock, Agricultural Research Council (ARC), Commonwealth Agricultural Bureaux (CAB).

Campling, R.E., Freer, M. and Balch, C.C., 1962. Factors affecting the voluntary intake of food by cows, 3. The effect of urea on the voluntary intake of oat straw. Br. J. Nutr., 16: 115-124.

Chesson, A. and Ørskov, E.R., 1984. Microbial degradation in the digestive tract. In: F. Sunstøl and E. Owen (eds.), Straw and Other Fibrous By-products as Feed. Developments in Animal and Veterinary Sciences, Elsevier, Amsterdam, pp. 305-339.

Chicco, C.F., Schultz, I.A., Schultz, E., Carnevalli, A.A. and Ammerman, C.B., 1972. Molasses-urea for restricted forage fed steers in the tropics. J. Anim. Sci., 35: 859-865.

Church, D.C. and Santos, A., 1981. Effect of graded levels of soybean meal and of a nonprotein nitrogen-molasses supplement on consumption and digestibility of wheat straw. J. Anim. Sci., 53: 1609-1615.

Coombe, J.B. and Tribe, D.E., 1962. The effects of urea supplements on the utilization of straw plus molasses diets by sheep. Austr. J. Agric. Res., 1: 70-91.

Dixon, R.M., 1984. Effect of various levels of molasses supplementation on intake of mature *Pennisetum purpureum* forage by growing cattle. Trop. Anim. Prod., 9: 30-34.

Ernst, A.J., Limpus, J.F. and O'Rourke, P.K., 1975. Effect of supplements of molasses and urea on intake and digestibility of native pasture hay by steers. Austr. J. Exp. Agric. Anim. Husb., 15: 451-455.

Ghebrehiwet, T., Ibrahim, M.N.M. and Schiere, J.B., 1988. Response of growing bulls to diets containing untreated or urea-treated rice straw with rice bran supplementation. Biol. Wastes, 25: 269-280.

ICAR, 1985. Terminal report 1967-1985. All India Coordinated Research Project 'Utilization of Agricultural By-products and Industrial Waste Materials for Evolving Economic Rations for Livestock'. Indian Council of Agricultural Research (ICAR), National Dairy Research Institute, Karnal, India.

Kunju, P.J.G., 1984. 'Urea Molasses Animal Lick' - An Approach to Increase the Utilization of Crop Residues. National Dairy Development Board, Anand, India, 4 pp.

Kunju, P.J.G., 1986. Urea molasses block lick: a feed supplement for ruminants. In: M.N.M. Ibrahim and J.B. Schiere (eds.), Rice Straw and Related Feeds in Ruminant Rations. Proc. Int. Workshop, Kandy, Sri Lanka, 24-28 March 1986. Straw Utilization Project (SUP), Kandy, Sri Lanka, pp. 261-274.

Leng, R.A., 1984a. Supplementation of tropical and subtropical pastures for ruminant production. In: F.C.M. Gilchrist and R.I. Mackie (eds.), Herbivore Nutrition in the Subtropics and Tropics. The Science Press, Graighall, South Africa, pp. 129-144.

Leng, R.A., 1984b. The potential of solidified molasses-based blocks for the correction of multi-nutritional deficiencies in buffaloes and other ruminants, fed low quality agro-industrial by-products. In: International Atomic Energy Agency (IAEA). Proceedings of FAO/IAEA meeting from 30 January-3 February 1984 in Manilla. IAEA, Vienna, pp. 130-150.

Loosli, J.K. and McDonald, I.W., 1968. Non-protein Nitrogen in the Nutrition of Ruminants. FAO, Rome.

Losada, H., Aranda, E., Ruiz, J. and Alderete, R., 1979. Effect of urea on voluntary intake and metabolic parameters in bulls fed sugarcane and molasses. Trop. Anim. Prod., 4: 168-171.

Manget Ram and Kunju, P.J.G., 1986. Effect of incorporating concentrate mixture with urea molasses mineral block feeding on ruminal metabolites and digesta flow rate in buffalo calves. Indian J. Anim. Nutr., 3: 244-248.

McLennan, S.R., Wright, G.S. and Blight, G.W., 1981. Effects of supplements of urea, molasses and sodium sulphate on the intake and liveweight of steers fed rice straw. Austr. J. Exp. Agric. Husb., 21: 367-370.

Ministry of Agriculture, 1957. Handbook for Farmers in South Africa, 3, Stock Farming and Pastures. Govt. Printer, Pretoria, South Africa.

Neric, S.P., Aquino, D.L., De la Cruz, P.C. and Ranjhan, S.K., 1985. Effect of urea-molasses-mineral block lick on the performance of caracows (swamp buffaloes). Indian J. Anim. Nutr., 2: 84-86.

Neter, J. and Wasserman, W., 1974. Applied Linear Statistical Models. Homewood, IL.

Pearce, J., 1973. Nutrient blocks for cattle. Agric. N. Ireland, 48: 288-290.

Preston, T.R. and Leng, R.A., 1984. Supplementation of diets based on fibrous residues and by-products. In: F. Sundstøl and E. Owen (eds.), Straw and Other Fibrous By-products as Feed. Elsevier, Amsterdam.

Saadullah, M., Haque, M. and Dolberg, F., 1981. Effectiveness of ammonification through urea in improving the feeding values of rice straw in ruminants. Trop. Anim. Prod., 6: 30-36.

Sansoucy, R., 1986. Manufacture of Molasses-Urea Blocks. World Anim. Rev., 57: 40-48.

Schiere, J.B. and Ibrahim, M.N.M., 1989. Feeding of Urea-ammonia Treated Rice Straw. A Compilation of Miscellaneous Reports Produced by the Straw Utilization Project. Pudoc, Wageningen, The Netherlands.

Snedecor, G.W. and Cochran, W.G., 1980. Statistical Methods, 7th edn. Iowa State University Press, Ames, IA.

Soetanto, H., Budi Sarono and Sulastri, 1987. Digestion and utilization of low quality roughages in goats and sheep as affected by access to molasses urea block in urea. Paper presented at the 7th Annual Meeting of AAFARR Network, Chiang Mai, Thailand, June 1987.

Sudana, I.B., 1985. Urea-molasses block for growing lambs on wheat straw basal diet. In: Efficient Animal Production for Asian Welfare. Proceedings of the 3rd AAAP Animal Science Congress, 6-10 May 1985, 2. Seoul, Korea.

Van Keulen, J. and Young, B.A., 1977. Evaluation of acid insoluble ash as a natural marker in ruminant digestibility studies. J. Anim. Sci., 44: 282.

Williams, S., 1984. Official Methods of Analysis of the Association of Official Analytical Chemists, 14th edn. A.D.A.C., Arlington, 1141 pp.

Chapter 3.3

OVERCOMING THE NUTRITIONAL LIMITATIONS OF RICE STRAW FOR RUMINANTS:

UREA AMMONIA UPGRADING OF STRAW AND SUPPLEMENTATION WITH RICE BRAN AND COCONUT CAKE FOR GROWING BULLS[1]

J.B. Schiere, V.R. Kumarasuntharam, V.J.H. Sewalt
and B. Brouwer

SUMMARY

Forty eight growing bulls of two breed types (red Sahiwal and white Kilari), fed rice straw, were allocated to nine treatment groups:

1. Control straw (CS)
2. Urea upgraded straw (UUS)
3. UUS + 0.25 kg coconut cake (CC)
4. UUS + 0.75 kg CC
5. UUS + 0.25 kg rice bran (RB)
6. UUS + 1.00 kg RB
7. UUS + 0.25 kg RB + 0.25 kg CC
8. UUS + 1.00 kg RB + 0.25 kg CC
9. CS + 1.00 kg RB + 0.25 kg CC

Liveweight gain was measured weekly during 15 weeks and tested in three analyses of variance. The results are:

Urea upgraded straw produced a liveweight gain 180 $g.d^{-1}$ higher ($P < 0.01$) than control straw. The groups supplemented with 0.25 kg coconut cake and 1.00 kg rice bran showed an increase of 100 $g.d^{-1}$ ($P < 0.05$) over the unsupplemented groups. No interaction between straw upgrading and supplementation was present ($P > 0.10$).

Both rice bran and coconut press cake, supplemented to upgraded straw at a level of 0.25 kg, did not increase liveweight gain ($P > 0.05$). However, 1.0 kg rice bran increased gain by 90 $g.d^{-1}$ ($P < 0.05$). A supplement of 0.75 kg coconut press cake to upgraded straw increased liveweight gain by 160 $g.d^{-1}$ compared with 0.25 kg or 0.00 kg coconut cake supplement ($P < 0.05$).

There were no significant differences between breed types ($P > 0.10$) or interactions between breed and the other two main treatments (upgrading and supplementation). It was concluded, that both urea upgrading and supplementation of rice straw increase animal performance. The effect of urea upgrading was the same for both supplemented and unsupplemented animals. There was no indication of a non-linear effect of supplements on growth.

---

[1] Published in Asian-Australian Journal of Animal Science, 1988, 1(4): 213-218

INTRODUCTION

The nutritional limitations of rice straw may be overcome by supplementation with concentrates, urea or green forage (Creek *et al.*, 1984; Preston and Leng, 1984; Ghebrehiwet *et al.*, 1988) or by upgrading of straw by chemical or physical treatment (Ibrahim, 1983), of which urea upgrading has proven to be very practical (Perdok *et al.*, 1982; Schiere *et al.*, 1988). In order to understand more about the effect of urea upgrading of straw *vs.* supplementation with concentrates, an experiment was conducted using coconut press cake and the relatively cheap rice bran fed as supplements to urea upgraded and untreated rice straw at different levels and combinations.

MATERIALS AND METHODS

Treatments

A group of forty eight growing bulls fed rice straw was divided into the following nine treatment groups:
1. Control straw (CS)
2. Urea upgraded straw (UUS)
3. UUS + 0.25 kg coconut cake (CC)
4. UUS + 0.75 kg CC
5. UUS + 0.25 kg rice bran (RB)
6. UUS + 1.00 kg RB
7. UUS + 0.25 kg RB + 0.25 kg CC
8. UUS + 1.00 kg RB + 0.25 kg CC
9. CS + 1.00 kg RB + 0.25 kg CC

The design of the experiment allowed for three treatment comparisons:
A. Control straw (1)
    Upgraded straw (2)
    Control straw + 0.25 kg CC + 1.00 kg RB (9)
    Upgraded straw + 0.25 kg CC + 1.00 kg RB (8)
B. Upgraded straw (2)
    Upgraded straw + 0.25 kg RB (5)
    Upgraded straw + 1.00 kg RB (6)
    Upgraded straw + 0.25 kg CC (3)
    Upgraded straw + 0.25 kg RB + 0.25 kg CC (7)
    Upgraded straw + 1.00 kg RB + 0.25 kg CC (8)
C. Upgraded straw (2)
    Upgraded straw + 0.25 kg CC (3)
    Upgraded straw + 0.75 kg CC (4)

A general objective was to determine whether the effect of concentrates is linear. In some cases, a stimulative effect of very small quantities of supplements on intake and liveweight gain have been reported (Saadullah, 1984; Leng and Van Houtert, 1986).

Animals

The 48 growing animals used consisted of two different breed types, red (mainly Sahiwal) and white (mainly Kilari). These were allocated to the treatments groups in such a way, that breed effects could be tested. Each treatment group contained five animals (three red and two white), except three groups, which contained six animals (four red and two white). All animals were young uncastrated bulls, weighing 80-160 kg (average 123 kg). The animals were housed and fed in groups. Before the experiment started, the animals were dewormed.

Feeds and feeding

The basal feed was rice straw obtained from village farmers. It was of unknown variety and cultivated under unknown fertiliser regimes. It was fed unchopped and *ad libitum,* either untreated or upgraded with 4% urea.

The upgraded straw was produced by addition of 4 kg urea in 100 l water to 100 kg airdry straw allowed to react for 9-11 days in large open heaps under a roof, not exposed to wind. After nine days, the upgraded straw was fed over the next three days. On the 12th day, a new lot of upgraded straw was started that had been made on the fourth day, etc.

Rice bran and coconut cake were fed in the morning and evening before the straw was offered. The rice bran was obtained from a local mill and was of the low quality generally available in Sri Lanka. For groups fed both rice bran and coconut cake, the concentrates were mixed together.

In addition to the experimental diets, all animals were fed 1 kg of fresh grass to supply vitamin A and simulate practical conditions. The grass was cut in the field irrespective of maturity and fed unchopped on top of the straw in the feed troughs. All animals were fed 30 g sodium sulphate, 20 g di-calcium phosphate and 50 g mineral mixture. The animals had free access to drinking water.

Measurements

The experiment lasted for 15 weeks, and liveweights were recorded before feeding at weekly intervals using a cattle scale. Liveweight gain was calculated by means of linear regression analysis (Snedecor and Cochran, 1980).

Statistical analysis

Liveweight gain was tested using analysis of variance (Snedecor and Cochran, 1980), in which initial body weight was added as a covariable. The Student-Newman-Keuls' test was used to check differences between treatment groups (Steel and Torrie, 1980). For comparison A, a three-way analysis was used with urea upgrading (control, upgraded), supplementation (unsupplemented, supplemented) and breed (red, white) as main effects. Comparison B was a three-way analysis with level of coconut cake (0, 0.25 kg) and level of rice bran (0, 0.25, 1.00 kg) and breed (red, white) as main effects. For comparison C, a two-way analysis was used with level of coconut cake (0, 0.25, 0.75 kg) and breed (red, white) as main effects. Comparison C was also combined with comparison B in a three-way analysis of variance with level of coconut cake (0, 0.25, 0.75 kg), level of rice bran (0, 0.25, 1.00 kg) and breed as main effects, to include more observations for the first two levels of coconut cake. In all analysis, interactions between main effects were tested.

RESULTS AND DISCUSSION

The results for comparisons A, B and C are summarised in Tables 1, 2 and 3. Means for treatment groups used in more than one comparison, differ slightly from one comparison to the other, due to the respective corrections for covariable effects.

Comparison A: Urea upgrading and supplementation with 1.00 kg rice bran plus 0.25 kg coconut cake.
Urea upgrading of straw increased liveweight gain by 182 g.d^{-1} (P < 0.01). Similar increases were found by Ghebrehiwet *et al.* (1988) and Schiere *et al.* (1989) who found liveweight gains on untreated straw of approximately -100 g.d^{-1} and on urea upgraded straw of +90 g.d^{-1}. Those levels are lower, however, than the levels found in this experiment, maybe due to a difference in the quality of the straw used. Tharmaraj *et al.* (1989) found a smaller improvement with upgrading (-121 g.d^{-1} on untreated straw and -4 g.d^{-1} on urea upgraded straw), maybe due to a less efficient treatment process in small open heaps as used in their experiment. The superiority of the urea upgraded straw is probably caused by a higher intake and digestibility of upgraded straw (Saadullah *et al.*, 1982; Chesson and Ørskov, 1984; Ghebrehiwet *et al.*, 1988; Schiere *et al.*, 1989). Doyle *et al.* (1986) found that in urea--ammonia upgrading about 75% of the increase in digestible organic matter intake was due to the supplementation with nitrogen and only a minor part to the chemical reaction of the ammonia released from urea with the cell wall component in straw.

Supplementation with 1.0 kg rice bran and 0.25 kg coconut cake to control straw or urea upgraded straw (Table 1) caused an increase of 98 g.d^{-1} (P < 0.05). No interaction between straw upgrading and supplementation was present (P > 0.10), indicating that the effect of urea upgrading is the same for supplemented and for unsupplemented groups, as also found by others (Ghebrehiwet *et al.*, 1988; Tharmaraj *et al.*, 1989). These improvements are somewhat lower than those found by Ghebrehiwet *et al.* (1988) who supplemented both untreated and urea upgraded straw with five levels of rice bran and found increases of 180 and 150 g.d^{-1} per kg rice bran addition for untreated and urea upgraded straw, respectively. The higher response to rice bran in their trial is probably due to a difference in rice bran quality. The quality of rice bran produced in Sri Lanka is highly variable, partially due to its variable ash content of 25-45% (Ibrahim, 1987).

Although the red animals performed better than the white animals in three of the four groups, no significant difference in favour of either type of animal emerged (P > 0.10). With the small number of animals used, interactions between breed and straw upgrading or between breed and supplement could not be detected (P > 0.10). No effect of initial weight (as a covariable) on liveweight gain was observed (P > 0.10).

Comparison B: Supplementation with three levels of rice bran and two levels of coconut cake to urea upgraded straw

Rice bran supplementation to upgraded straw at a level of 1.0 kg significantly ($P < 0.05$) increased the liveweight gain with 100 g.d^{-1}. This resulted in a gain of 254 g.d^{-1} (Table 2), which is the same growth as found by Ghebrehiwet *et al.* (1988) for Sahiwal crosses on urea upgraded straw supplemented with 1.0 kg rice bran. The effects of 0.25 kg coconut cake or 0.25 kg rice bran were not significant ($P > 0.05$). Initial body weight affected liveweight gain significantly ($P < 0.05$), due to a high variation in initial weight within some of the treatment groups. No breed effect and no interactions were present ($P > 0.10$).

Table 1.   Effect of urea upgrading and supplementation with 1.00 kg rice bran plus 0.25 Kg coconut cake on liveweight gain of growing bulls of 2 breed types[1]

	Control straw		Upgraded straw	
	without supplement	with supplement	without supplement	with supplement
Liveweight gain (g.d^{-1})				
Red animals	22	98	179	235
White animals	-98	30	108	318
All animals[2,3]	-26[a]	68[ab]	146[b]	256[c]

[1]abc   Values with the same superscripts are not significantly different ($P > 0.05$). Breed effects were not significant ($P > 0.10$).

[2]   These average values are corrected for covariable effect of initial body weight.

[3]   Breed effects were not significant ($P > 0.10$).

Table 2.   Effect of supplementation with rice bran and coconut cake to urea upgraded straw on liveweight gain of growing bulls of 2 breed types[1]

	Level of coconut cake (kg fresh matter)					
	0.00			0.25		
Level of rice bran (kg)	0.00	0.25	1.00	0.00	0.25	1.00
Liveweight gain (g.d^{-1})						
Red animals	179	169	274	206	142	235
White animals	108	196	196	102	246	318
All animals[2,3]	154[a]	154[a]	254[b]	145[a]	208[a]	271[b]

[1]abc   Values with the same superscripts are not significantly different ($P > 0.05$). Breed effects were not significant ($P > 0.10$).

[2]   These average values are corrected for covariable effect of initial body weight.

[3]   Breed effects were not significant ($P > 0.10$).

Comparison C: Supplementation with three levels of coconut cake to urea upgraded straw.

Table 3 shows that liveweight gain at a supplementation level of 0.75 kg coconut cake is approximately 160 g.d^{-1} higher than at 0.25 kg or 0.00 kg supplement. However, in the two-way analysis of variance coconut cake supplementation did not affect liveweight gain significantly ($P > 0.05$), due to low animal numbers per class. At the higher animal numbers included in the three-way analysis (including the treatment groups involved in comparison B), the effect of 0.75 kg coconut cake became significant ($P < 0.05$). In an experiment with growing Sahiwals, Perdok *et al.* (1984) found a similar increase of 150 g.d^{-1} ($P < 0.05$), when urea upgraded straw was supplemented with 0.6 kg (dm) coconut cake. No breed effect or interaction between breed and coconut cake level was observed at these numbers of animals ($P > 0.10$). Inclusion of initial weight as a covariable did not result in a significant covariable effect ($P > 0.10$).

Table 3.   Effect of supplementation with coconut cake to urea upgraded straw on liveweight gain of growing bulls of two breed types[1]

	Level of coconut cake (kg fresh matter)		
	0.00	0.25	0.75
Liveweight gain (g.d^{-1})			
Red animals	179	206	315
White animals	108	103	331
All animals[2,3]	154[a]	146[a]	322[b]

[1] a,b Values with the same superscripts are not significantly different ($P > 0.05$).
[2]   These average values are corrected for covariable effect of initial body weight.
[3]   Breed effects were not significant ($P > 0.10$).

CONCLUSION

This experiment shows clearly that animal performance on rice straw can be increased by either upgrading or supplementing straw, or by a combination of these. In this experiment, the supplements consisted of rice bran and coconut cake at several levels and combinations. Non-linear effects of small amounts of supplements could not be indicated. Such non-linear effects might be expected, considering the non-linear effect as found by Saadullah (1984) in the case of fish meal and considering the effect of small quantities (50-100 g.d^{-1}) of protein meal on liveweight gain (Van Houtert and Leng, 1986). In this experiment, small supplements of both rice bran and coconut cake did not increase liveweight gain significantly. The absence of interaction between straw upgrading and supplementation, as also found by others (Schiere *et al.*, 1985a; Ghebrehiwet *et al.*, 1988; Tharmaraj *et al.*, 1989), indicates that the effect of urea upgrading is the same for animals that are supplemented or not supplemented. In this experiment animals on urea upgraded straw alone grew at a rate of 150 g.d^{-1}, while for animals on untreated straw, rice bran should constitute almost 50% of the ration to obtain the same growth rate. At such high levels of concentrates, problems can arise regarding the intake of straw. The choice between the alternatives has to be based on economics. Ration calculations have shown that feeding urea upgraded straw is profitable at higher levels of production or when concentrates are expensive (Schiere *et al.*, 1985b; Nell *et al.*, 1986).

ACKNOWLEDGEMENTS

This paper reports an experiment conducted in 1983 by the Straw Utilization Project, which was based on collaboration of the Directorate General of International Cooperation of the Netherlands with the National Livestock Development Board, Sri Lanka, the University of Peradeniya, Sri Lanka and the Mahaweli Authority of Sri Lanka. Thanks are due to H.B. Perdok (Straw Treatment Project), to M.C.N. Jayasuriya (University of Peradeniya), to the manager of Niravia farm (MASL), Mr. Canagasaby and his farm staff, to the consultant of MASL-DAP, Mr. C. de Saram and to J. Smit and R. Van Der Hoek for their work during the implementation of the experiment.

REFERENCES

Chesson, A. and E.R. Ørskov. 1984. Microbial degradation in the digestive tract. In: Sundstøl, F. and E. Owen (eds.), Straw and other fibrous by-products as feed. Developments in Animal and Veterinary Sciences, 14. Elsevier, Amsterdam, pp. 305-339.

Creek, M.J., T.J. Barker and W.A. Hargus. 1984. The development of a new technology in an ancient land. World Animal Review 51: 12-20.

Doyle, P.T., C. Devendra and G.R. Pearce. 1986. Rice straw as a feed for ruminants. International Development Program of Australian Universities and Colleges (IDP), Canberra, 117 pp.

Ghebrehiwet, T., M.N.M. Ibrahim and J.B. Schiere. 1988. Response of growing bulls to diets containing untreated straw or urea-treated rice straw with rice bran supplementation. Biological Wastes, 25: 269-280.

Ibrahim, M.N.M. 1983. Physical, chemical, physico-chemical and biological treatment of crop residues. In: Pearce, G. (ed.), The utilization of fibrous agricultural residues, Proc. of a seminar in Los Banos, Philippines. Australian Govt. Publ. Service, Canberra, pp. 53-68.

Ibrahim, M.N.M. 1987. Rice bran as a supplement for straw based rations. In: Dixon, R.M. (ed.), Ruminant feeding systems utilizing agricultural residues-1986. Proceedings of the 6th annual workshop of the AAFARR Network held in the University of the Philippines at Los Banos, 1-3 April, 1986. IDP, Canberra, pp. 139-146.

Ibrahim, M.N.M. and J.B. Schiere. 1986. Rice straw and related feeds in ruminant rations. Proceedings of an international workshop held in Kandy, Sri Lanka, 24-28th March, 1986. Straw Utilization Project, Kandy, Sri Lanka.

Nell, A.J., J.B. Schiere and M.N.M. Ibrahim. 1986. Economic evaluation of urea-ammonia treated straw as cattle feed. In: Ibrahim, M.N.M. and J.B. Schiere (eds.), Rice straw and related feeds in ruminant rations, Proceedings of an international workshop held in Kandy, Sri Lanka, 24-28 March, 1986. Straw Utilization Project, Kandy, Sri Lanka, pp. 164-170.

Perdok, H.B., M. Thamotharam, J.J. Blom, H. Van Den Born and H. Van Veluw. 1982. Practical experiences with urea ensiled straw in Sri Lanka. In: Preston, T.R., C.H. Davis, F. Dolberg, M. Haque, M. and M. Saadullah (eds.), Maximum Livestock Production from Minimum Land. Proceedings of the third seminar held in Bangladesh, 13-18th Feb. pp. 123-134.

Perdok, H.B., G.S. Muttetuwegama, G.A. Kaasschieter, H.M. Boon, N.M. Van Wageningen, V. Arumugam, M.G.F.A. Linders and M.C.N. Jayasuriya. 1984. Production responses of lactating or growing ruminants fed urea-ammonia treated paddy straw with or without supplements. In: Doyle, P.T. (ed.), The Utilization of Fibrous Agricultural Residues for Animal Feeds. Proceedings of the third annual workshop of the

AAFARR Network. School of Agriculture and Forestry, University of Melbourne, Parkville, Victoria, pp. 213-230.

Preston, T.R. and R.A. Leng. 1984. Supplementation of diets based on fibrous residues and by-products. In: Sundstøl, R.F. and E. Owen (eds.), Straw and other fibrous by-products as feed. Developments in Animal and Veterinary Sciences, 14. Elsevier, Amsterdam, pp. 373-413.

Saadullah, M. 1984. Ph.D. Thesis, Royal Veterinary and Agricultural University, Copenhagen, Denmark.

Saadullah, M., M. Haque and F. Dolberg. 1982. Treated and untreated rice straw for growing cattle. Tropical Animal Production 7: 20-25.

Schiere, J.B. and M.N.M. Ibrahim. 1985. Recent research and extension on rice straw feeding in Sri Lanka: A review. In: Doyle, P.T. (ed.), The utilization of fibrous agricultural residues as animal feeds. Proc. of the fourth annual workshop of the AAFARR Network, Khon Kaen, Thailand, 10-14 April, 1984, IDP, Canberra, pp. 116-126.

Schiere, J.B., M.N.M. Ibrahim and A. De Rond, 1985a. Supplementation of urea-treated rice straw. In: Wanapat, M. and C. Devendra (eds.), Relevance of crop residues as animal feeds in developing countries. Proc. of an international workshop in Khon Kaen, Thailand, pp. 273-300.

Schiere, J.B., A.J. Nell and M.N.M. Ibrahim. 1985b. Approaches to determine the economies of feeding urea-ammonia treated straw, untreated straw or urea supplemented straw. In: Doyle, P.T. (ed.), The utilization of fibrous agricultural residues as animal feeds. Proc. of the fourth annual workshop of the AAFARR Network, Khon Kaen, Thailand, 10-14 April, 1984, IDP, Canberra, pp. 127-131.

Schiere, J.B., A.J. Nell and M.N.M. Ibrahim. 1988. Feeding of urea-ammonia treated rice straw. World Animal Review 65: 31-42.

Schiere, J.B., M.N.M. Ibrahim, V.J.H. Sewalt and G. Zemmelink. 1989. Response of growing cattle given rice straw to lickblocks containing urea and molasses. Animal Feed Science and Technology, 26: 179-189.

Snedecor, G.W. and W.G. Cochran. 1980. Statistical Methods (7th edition). Iowa State University Press, Ames, Iowa, 459 pp.

Steel, R.G.D. and J.H. Torrie. 1980. Principles and procedures of statistics. A biometrical approach (2nd ed.). McGraw-Hill, Singapore.

Tharmaraj, J., R. Van Der Hoek, V.J.H. Sewalt and J.B. Schiere. 1989. Overcoming the nutritional limitations of rice straw for ruminants. 4. Urea ammonia treatment and supplementation with *Gliricidia maculata* for growing sahiwal bulls. Asian-Australian Journal of Animal Sciences, 2 (2): 85-90.

Van Houtert, M. and R.A. Leng. 1986. Strategic supplements to increase the efficiency of utilization of rice straw by ruminants. In: Rice straw and related feeds in ruminant rations. Proceedings of an international workshop held in Kandy, Sri Lanka, 24-28th March, 1986. SUP, Kandy, Sri Lanka, pp. 282-284.

Chapter 3.4.

RESPONSE OF GROWING BULLS TO DIETS CONTAINING UNTREATED OR UREA-TREATED RICE STRAW WITH RICE BRAN SUPPLEMENTATION[1]

T. Ghebrehiwet, M.N.M. Ibrahim and J.B. Schiere

SUMMARY

Untreated or urea-treated (4% w/w) rice straw supplemented with five levels of rice bran (0-1.6 kg/day) was given to 74 growing cross-bred bulls. Urea treatment significantly (P < 0.001) improved feed dry matter intake at all levels of rice bran supplementation. Feeding treated straw alone increased dry matter intakes up to 30%. This could be partly due to the significantly higher (P < 0.01) organic matter digestibility of the treated straw (59 *vs.* 53%). At all levels of supplementation, treated straw gave significantly higher (P < 0.001) liveweight gains than untreated straw. Animals fed on treated straw alone gained 93 g/day, whereas those fed on untreated straw lost 123 g/day. Treating straw with urea economises on rice bran supplementation or results in higher liveweight gain for the same level of supplementation.

[1] Published in Biological Wastes, 1988, 25(4): 269-280.

INTRODUCTION

Cereal straws are basically energy feeds with low protein contents and an unbalanced mineral composition. However, from such feeds the energy available to ruminants is low owing to their low digestibility (40-50%), which can indirectly lead to relatively low voluntary intake (Jackson, 1977). The nutritive value of rice straw could be improved by treatment and/or supplementation. While in the developed world more emphasis is placed on the use of ammonia (Sundstol *et al.*, 1978; Creek *et al.*, 1984), Asian scientists have considered the use of urea-ammonia treatment (Perdok *et al.*, 1982, Verma, 1983: Saadullah, 1984; Ibrahim and Schiere, 1985; Wanapat, 1985). The above researches have clearly shown that urea treatment increases the feeding value of straw by raising digestibility and intake. Also the extra nitrogen supply is an important advantage in using ammonia or urea. It has also been demonstrated that treatment reduces the need for concentrate supplementation (Creek *et al.*, 1984; Schiere *et al.*, 1985a). Nevertheless, concentrates may still be needed for higher levels of production to balance specific deficiencies, or concentrate supplementation may be cheaper than chemical treatment to attain a given level of production. Rice bran is the most widely available and the cheapest among the concentrate feeds in Sri Lanka, but its nutritive value is highly variable (Leelawardane, 1985, unpublished). The experiment reported in this paper was designed to obtain more information on the difference in animal response to rice bran supplementation with untreated or urea-treated rice straw.

METHODS

Animals, diets and experimental design

Seventy-four bull calves (indigenous x Sahiwal crosses) 12-16 months of age and weighing 100-160 kg were selected from a grazing herd of 300 animals. The selected animals were de-wormed, eartagged and classified into weight groups (based on the average of three consecutive days of weighing). The animals were then randomly allocated to ten groups, each consisting of seven or eight animals.

The animals were housed, tethered and equipped with individual feed troughs. The animals were fed unchopped untreated or urea-treated rice straw, with or without rice bran as a supplement. The urea-treated rice straw was prepared by mixing straw with urea solution (4 kg urea dissolved in 100 litres water/100 kg straw) and storing it in cement-lined pits covered with polythene sheets for at least 7 days. Each pit accommodated one week's feed requirement and the treated straw was offered unaerated.

The ten diets formed a 5 x 2 factorial design and consisted of untreated straw (US) or treated straw (TS), supplemented with 0, 400, 800, 1200 or 1600 g rice bran. In addition to straw and rice bran, each animal was given 1 kg fresh grass (*Brachiaria ruziziensis*-25% dry matter) and 100 g mineral mixture (34 g NaCl, 33 g Na_2SO_4, 33 g mineral premix) daily. Rice bran was offered separately in wooden boxes for each animal according to the allocated level. Of the mineral mixture, salt (NaCl) was mixed with the rice bran (supplemented groups) or mixed with the straws (US_0 and TS_0 groups). For all treatment groups, the sodium sulphate and the mineral premix were mixed with the straws. All animals had free access to drinking water.

Feed intake and liveweight gain measurements

The experiment lasted for 14 weeks, which included an adaptation period of 2 weeks, pre-experimental period of 3 weeks and an experimental period of 9 weeks. The straw given during the experimental period came from three batches, of which two were freshly harvested. During the first 7 weeks of the experimental period, daily group dry matter intakes, straw dry matter intakes, rice bran dry matter intakes and weekly liveweights were measured. During the last 2 weeks of the experimental period, digestibility of five selected diets was measured. Although each animal had an individual feeding trough, owing to the large number of animals involved straw dry matter intakes were measured on a treatment group basis. Straw (US/TS) was weighed for a group and divided among the animals in that group, being offered *ad libitum* (15-20% in excess of previous day's intake). Straw refusals were also collected and weighed for the group. Composite samples of the straw offered and refused were taken once a week for each treatment group and oven dried at 70°C for 30 h for dry matter determination. Depending on the level allocated to each treatment group, rice bran was offered to each animal in 400 g amounts. Weights of rice bran refused by each animal were recorded and then the refusals of all animals were bulked, mixed well and a representative subsample was taken for dry matter determination. Due to the limitations of labour and the time involved in weighing the animals in one day, the animals were grouped into two and weighed weekly on successive days. The animals were weighed every Monday/Tuesday and before offering feed. The daily liveweight gain was estimated by linear regression of liveweight on time.

Digestibility measurement

During the last 2 weeks of the experimental period fresh grass and mineral feeding were withdrawn and five animals from each of treatment groups of the US_0, US_2, US_4, TS_0 and TS_2 were selected for estimation of digestibility. During a collection period of 10 days their individual straw and rice bran dry matter intakes were measured daily. Also their faecal outputs were collected manually and each 24-h output was mixed thoroughly and a 10% subsample was dried at 105°C for 48 h for dry matter determination. At the end of the collection period, the daily samples of straw and rice bran were mixed well and ground through a 1-mm sieve. Also the dried faecal samples were bulked separately for each animal and ground through a 1-mm sieve. Dry matter and ash were determined (AOAC, 1970) in these feed and faecal samples. Samples of untreated straw, treated straw (as offered), rice straw and grass were analyzed for nitrogen by the macro Kjeldahl method of the Association of Official Analytical Chemists (1980). Organic matter digestibility *in vitro* of the rice bran was also determined by the method of Tilley and Terry (1963).

Statistical analyses

The results were subjected to analyses of variance (SPSS-ANOVA). The data on intake of straw were analysed to test the effect of treatment of straw, week, rice bran intake and mean liveweight. Also the individual data for liveweight gain as affected by treatment, rice bran intake, initial liveweight and weeks were analysed.

RESULTS

The crude protein content (%DM) of the untreated straw, treated straw, grass and rice bran used in the different rations were 4.5%, 10.0%, 11.0% and 9.0%, respectively. Group average dry matter intakes of straw, rice bran and grass are presented in Table 1. Urea treatment of straw significantly increased (P<0.001) its dry matter intake. The unsupplemented treated straw (TS_0) intake was 34% higher than the unsupplemented untreated straw (US_0) intake. The effects due to week and the treatment x week interaction were also significant (P<0.01).

In the supplemented groups the increase in dry matter intake of treated straw over untreated straw ranged from 28% to 52% (Table 2). The higher percentages should not be overemphasised because the US groups had higher rice bran intakes than the TS groups. This is shown by the decrease in the percentage increase in total dry matter intake of TS over US groups with the increase in quantity of rice bran offered.

Rice bran significantly decreased (P<0.001) straw intakes, and the decrease was more pronounced with untreated straw (7-17%) than with treated straw (2-10%). This was mainly due to the higher intake of rice bran in the US groups. Rice bran dry matter intake increased in both US and TS groups with the increase in the level of rice bran offered, but the rice bran intakes of animals given TS were lower (6-19%) than those of corresponding groups given US (10-31%). The relationships between the amount of rice bran offered and consumed by the two treatment groups are shown in Fig. 1. At higher levels of supplementation, the deviation of the intake line from the offered line is greater in TS groups than in the US groups.

The effect of rice bran intake on total intake is shown in Fig. 2. Both treatment and rice bran significantly affected (P<0.001) the total dry matter intake. When the dry matter intake was expressed in terms of metabolic body weight (g/kg $W^{-0.75}$/day), both treatment and rice bran intake showed significant effects (P<0.001) on straw dry matter intake.

Organic matter digestibility and dry matter intakes of five of the diets are presented in Table 3. Treatment of straw significantly increased (P<0.005) the organic matter digestibility. Rice bran supplementation also showed significant effects on organic matter digestibility (P<0.005). The organic matter digestibility *in vitro* of the rice bran used in this experiment was 50.2%. Supplementation of untreated straw with rice bran decreased the digestibility of the diet, whereas with treated straw marginal increases were shown.

Table 1.   Liveweight of animals, crude protein (CP) content of diet and the dry matter intakes of animals fed untreated or treated straw diets (group averages)

	Untreated straw					Treated straw				
	Rice bran offered (g)					Rice bran offered (g)				
	0	400	800	1200	1600	0	400	800	1200	1600
Number of animals	8	7	8	7	7	8	7	8	7	7
Mean liveweight (kg)	126.0	127.1	122.5	131.7	127.3	133.8	130.8	134.6	135.3	129.1
	(0.95)	(0.68)	(0.46)	(0.68)	(0.60)	(0.67)	(0.44)	(1.48)	(1.59)	(1.10)
CP content of diet[a] (%DM)	5.1	5.5	5.8	6.0	6.5	10.1	9.9	9.9	9.7	9.8
Dry matter intake (kg/animal/day)										
Straw	2.90	2.70	2.50	2.90	2.40	3.90	3.60	3.80	3.70	3.50
Rice bran	-	0.32	0.70	1.00	1.20	-	0.25	0.52	0.90	0.67
Grass	0.25	0.25	0.25	0.25	0.25	0.24	0.21	0.23	0.23	0.23
Total	3.15	3.27	3.45	4.15	3.85	4.14	4.06	4.55	4.83	4.40
Dry matter intake (kg/100 kg LW/day)										
Straw	2.30	2.10	2.00	2.20	1.90	2.90	2.80	2.80	2.80	2.70
Rice bran	-	0.25	0.57	0.76	0.94	-	0.20	0.39	0.68	0.52
Grass	0.20	0.20	0.20	0.19	0.20	0.18	0.16	0.17	0.17	0.21
Total	2.50	2.55	2.77	3.15	3.04	3.18	3.16	3.26	3.65	3.43
	(0.13)	(0.07)	(0.06)	(0.08)	(0.12)	(0.06)	(0.04)	(0.08)	(0.07)	(0.08)

Figures in parentheses are standard error.
[a] Calculated from individual intakes of straw, rice bran and grass.

Table 2.   Percentage increase in straw and total dry matter intake of treated straw (TS)
                over untreated straw (US) groups

Comparative treatment groups	Percentage increase in dry matter intake[a]	
	Straw	Total
$TS_0 > US_0$	34	31
$TS_1 > US_1$	33	24
$TS_2 > US_2$	52	32
$TS_3 > US_3$	28	16
$TS_4 > US_4$	46	14

[a] $\{(TS-US)/US\} \times 100-100$

Table 3.   Dry matter intakes and organic matter digestibility of some untreated and
                treated straw rations

Treatment group	Dry matter intake		Organic matter digestibility (%)
	Straw (kg/day)	Rice bran (g/day)	
US_0	3.1±0.6	-	53.4±1.5
US_2	3.0±0.5	602±80	49.9±3.3
US_4	3.0±0.3	1330±90	47.4±1.5
TS_0	3.9±0.5	-	59.4±5.9
TS_2	4.5±0.2	578±96	64.8±3.1

± = Standard deviation
- = Not offered

Feeding on treated straw significantly increased ($P < 0.001$) the daily liveweight gains. The
liveweight gains of animals not given rice bran were -123 g/day for untreated straw and 93
g/day for treated straw. The animals in the treated straw group gained on average 185 ±
87 g/day, while those in the untreated straw group lost on average 5 ± 112 g/day (Fig. 3).
The effects due to weeks ($P < 0.001$) and due to treatment x week interaction ($P < 0.05$)
were also significant. The relationship between daily liveweight gain and rice bran dry
matter intake during the whole experimental period is shown in Fig. 3. Rice bran
supplementation significantly increased ($P < 0.001$) the daily liveweight gain of animals
given both untreated and treated straw. From the slopes of the regression lines it is evident
that there is no significant difference in the increase in liveweight gain with the increase
in rice bran intake between the untreated and treated straw groups.

Figure 1. Relationship between amount of rice bran offered and consumed.

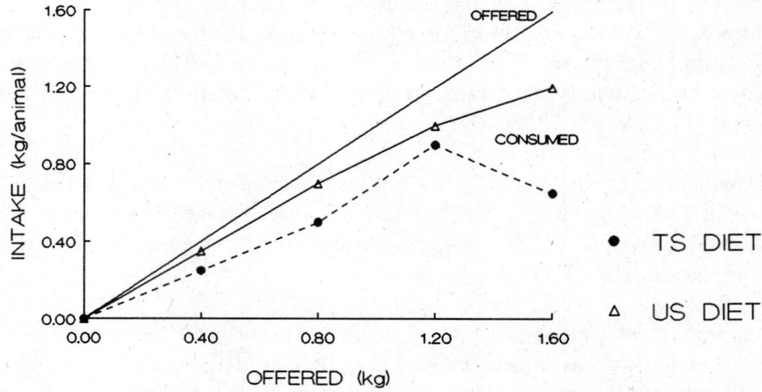

Figure 2. Total dry matter intake (TDMI) as affected by treatment and rice bran intake.

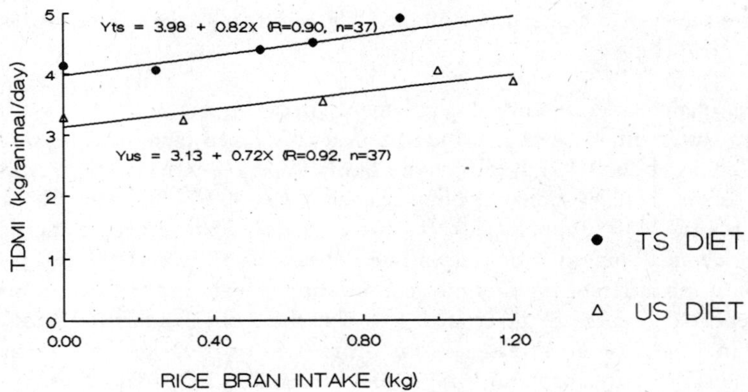

Figure 3. Relationship between daily liveweight gain and rice bran intake (according to regression).

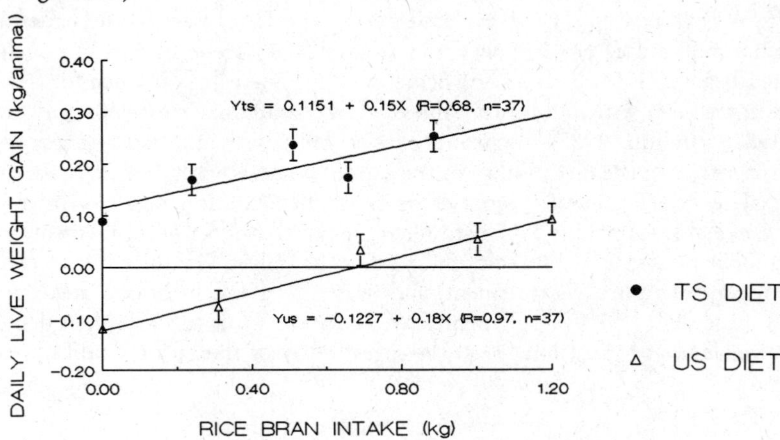

DISCUSSION

Urea treatment increased the dry matter intake of the basal diet (straw). The treatment groups showed differences in straw dry matter intakes due to treatment and the level of rice bran supplementation. The differences in straw intakes at higher levels of supplementation are mainly due to the greater substitution effect of rice bran on untreated than on treated straw.

Urea treatment increased the intake by 34%, which agrees with the 30% increase reported by Creek *et al.* (1984). Straw dry matter intakes as a percentage of liveweight for untreated (1.9-2.3%) and treated (2.7-2.9%) straw are also similar to those reported by other workers (Boon, 1983; Saadullah, 1984).

The higher intakes of treated straw required less concentrate supplementation to achieve the same liveweight gain as for untreated straw. In a similar experiment in Sri Lanka with gliricidia (*Gliricidia maculata*) supplementation, animals fed on treated straw consumed less gliricidia than animals fed on untreated straw (Straw Utilization Project, unpublished data). Creek *et al.* (1984) reported that treatment of straw saves concentrate feeding and higher savings are achieved at higher straw intakes. The results of the experiment presented in this paper confirm this.

The type, quality and amount of supplement affect the intake of the basal feed and the total dry matter intake. The rice bran used in this experiment was of poor to medium quality, having about 35% ash (dry matter basis), 9% ether extract and digestibility *in vitro* of 50%. In Sri Lanka, similar quality rice bran has 40% TDN and 6% crude protein (Leelawardana, 1985, unpublished). In this experiment the expected increases in straw intakes with small amounts of concentrate did not occur, in fact the reverse was found. Similarly, using different types of concentrate supplements and their combinations, Boon (1983) reported a decrease in treated straw intakes. But Saadullah (1984) reported an increase in intake of urea-treated straw (from 1.64 to 1.92 kg/day) when a fish-meal supplement (55% crude protein) was increased from 0 to 250 g/day. This clearly indicates that a poor-quality supplement only replaces straw instead of stimulating its consumption.

Treatment of rice straw with urea significantly increased its organic matter digestibility by 6 units (53.4% to 59.4%). The straw used during the latter part of the experiment was freshly harvested and of good quality and a digestibility value of 53% for such straw is acceptable (Ibrahim, 1985; Ibrahim and Schiere, 1985). The value obtained for treated straw is also in agreement with published values for urea-ammonia treatment (Jayasuriya and Perera, 1982; Ibrahim, 1985). Inclusion of rice bran with untreated straw resulted in marginal decreases in the digestibility of the ration, whereas with treated straw the reverse was true. The interaction between rice bran supplementation and treatment is rather difficult to explain. A possible explanation for the positive effect of supplement on TS may be due to the increased activity and population of the cellulolytic microbes in the presence of extra nitrogen (applied via treatment) and readily available carbohydrates (supplied via rice bran). Saadullah (1984) supplemented treated straw with fish-meal and demonstrated that at lower levels of inclusion (50 g) the digestibility of the ration could be increased.

The better nutritive value of treated straw (higher intake and digestibility) resulted in positive liveweight gains even without supplementation (93 g/animal/day), while feeding on untreated straw resulted in loss of weight (-123 g/animal/day). Nurazzamal *et al.* (1981) reported daily gains of 109 g/animal and losses of 149 g/animal for cattle fed on treated and untreated straw, respectively.

The effect of supplementation on liveweight gain depends on the type (quality) and the amount of supplement offered. In the study reported here, the voluntary rice bran consumption was higher with untreated straw than with treated straw. As such, the untreated-straw groups had a higher liveweight gain change with the increase in level of rice bran offered than the treated-straw groups, but this response was not significantly different. Creek *et al.* (1984) reported a positive response to concentrate supplementation with both untreated and ammonia-treated straw. But they found significantly higher (P < 0.01) liveweight gain response to concentrate supplementation of untreated-straw groups than the treated-straw groups. The differences in response to untreated and treated straw in the two experiments could be due to the type of treatment (ammonia gas *vs.* urea-ammonia), quality of straw, and the quality and quantity of concentrate supplements used.

Davis (1982, unpublished) showed that when cattle were fed on treated rice straw with and without an oil-cake supplement, the unsupplemented group gained at the rate of 84 g/day, which is similar to the gains reported in our study (93 g/day). Perdok *et al.* (1982) gave untreated and urea-treated rice straw with 500 g of concentrate supplement (type of concentrate not known) and reported liveweight gains of 73 and 346 g/day, respectively. In our experiment, supplementation of untreated and treated rice straw with about 500 g rice bran resulted in daily gains of 33 and 240 g, respectively. These differences could be partly explained by the above-mentioned reasons.

Quality of rice bran found in Sri Lanka is highly variable and, for example, the ash content could range from 25% to 45% (Leelawardana, 1985, unpublished). Nevertheless, the results of the experiments reported here clearly indicate that animal performance could be increased by using rice bran as a supplement and/or by giving urea-treated straw. Animals fed on untreated straw with 500 g rice bran could maintain weight, but to obtain daily gains around 100 g rice bran should constitute at least 30% of the ration. In contrast, about 100 g daily gain could be obtained by giving urea-treated straw alone. The choice between these alternatives will be an economic issue.

Such economic calculations should take into consideration the type (quality) and cost of supplement, cost of untreated and treated straw, expected levels of production and market price of animal products (milk/meat). From calculations based on the findings of this experiment (to obtain 100 g gain/day) and the current market price of rice straw and meat in Sri Lanka, it is more economical to feed untreated straw supplemented with rice bran than to feed treated straw. Similar ration calculations (Schiere *et al.*, 1985b) have shown that feeding treated straw is profitable at higher levels of production or when concentrates are expensive.

ACKNOWLEDGEMENTS

This paper is part of the work undertaken by the first author for his doctoral thesis (Ir.) at the Department of Tropical Animal Production, Wageningen Agricultural University. Funds were provided by the Straw Utilization Project, Sri Lanka. The authors wish to thank the National Livestock Development Board for providing facilities (Galpokuna farm) to conduct this experiment and Mr P. Siritunge (Manager) for all the assistance given. The authors also wish to thank the laboratory staff of the Animal Nutrition Laboratory, Department of Animal Science, and Mr B. Brouwer (Department of Tropical Animal Production) for assisting in the statistical analyses of the data.

REFERENCES

AOAC (1970). *Official Methods of Analyses*, 11th edn. Association of Official Analytical Chemists, Washington, DC.

AOAC (1980). *Official Methods of Analyses*, 13th edn. Association of Official Analytical Chemists, Washington, DC.

Boon, H.M. (1983). The effect of supplemented urea ensiled paddy straw on growth. Dept Trop. Anim. Husb., National Agric. Univ., Wageningen. Onderzoeksverslagen, Sri Lanka (unpublished), 23 pp.

Creek, M.J., Barker, T.J. and Hargus, W.A. (1984). The development of a new technology in an ancient land. *World Anim. Rev.*, **51**, 12-20.

Ibrahim, M.N.M. (1985). Effects of *Gliricidia maculata* leaves on the duration of urea-ammonia treatment of rice straw. In *The Utilization of Fibrous Agricultural Residues as Animal Feeds*, (ed). P.T. Doyle. University of Melbourne Printing Services, Parkville, pp. 77-80.

Ibrahim, M.N.M. and Schiere, J.B. (1985). Factors affecting urea-ammonia treatment. In *Relevance of Crop Residues as Animal Feeds in Developing Countries*, (eds.) M. Wanapat and C. Devendra. Funny Press, Bangkok, pp. 176-87.

Jackson, M.G. (1977). Review article: The alkali treatment of straw. *Anim. Feed Sci. Technol.*, **2**, 105-30.

Jayasuriya, M.C.N. and Perera, H.G.D. (1982). Urea-ammonia treatment of rice straw to improve its nutritive value for ruminants. *Agric. Wastes*, **4**, 143-50.

Nurazzamal Khan, A.K.M. and Davis, C.H. (1981). Effect of treating paddy straw with ammonia (generated from urea) on the performance of local and crossbred lactating cows. In *Maximum Livestock Production from Minimum Land*, (eds.) M.G. Jackson, F. Dolberg, C.H. Davis, M. Haque and M. Saadullah, Joyedebpur. Bangladesh, pp. 168-80.

Perdok, H.B., Thamotharam, M., Blom, J.J., Van Den Born, H. and Van Veluw, C. (1982). Practical experiences with urea-ensiled straw in Sri Lanka. In *Maximum Livestock Production from Minimum Land*, (eds.) T.R. Preston, C.H. Davis, F. Dolberg. M. Haque and M. Saadullah. Joydebpur, Bangladesh, pp. 123-34.

Saadullah, M. (1984). Supplementing ammoniated rice straw for native cattle in Bangladesh. Ph.D. Thesis, Institute of Animal Science, The Royal Veterinary and Agricultural University, Copenhagen, Denmark.

Schiere, J.B., Ibrahim, M.N.M. and Rond, A. De (1985a). Supplementation of urea-treated rice straw. In *Relevance of Crop Residues as Animal Feeds in Developing Countries*, (eds.) M. Wanapat and C. Devendra. Funny Press, Bangkok, pp. 273-300.

Schiere, J.B., Nell, A.J. and Ibrahim, M.N.M. (1985b). Approaches to determine the economics of feeding urea ammonia treated straw in urea supplemented straw. In *The Utilization of Fibrous Agricultural Residues as Animal Feed*, (ed.) P.T. Doyle. IDP, Canberra, pp. 127-31.

Sundstol, F., Coxworth, E. and Mowat, D.N. (1978). Improving the nutritive value of straw and other low quality roughages by treatment with ammonia. *World Anim. Rev.*, **26**, 13-21.

Tilley, J.M.A. and Terry, R.A. (1963). A two-stage technique for *in vitro* digestion of forage crops. *J. Br. Grassl. Soc.*, **18**, 104-11.

Verma, M.L. (1983). Practical aspects of treatment of crop residues. In *The Utilization of Fibrous Agricultural Residues*, (ed.) G.R. Pearce. Australian Government Publishing Service, Canberra, ACT, pp. 85-99.

Wanapat, M. (1985). Improving rice straw quality as ruminant feed by urea treatment in Thailand. In *Relevance of Crop Residues as Animal Feeds in Developing Countries*, (eds.) M. Wanapat and C. Devendra. Funny Press, Bangkok, pp. 147-75.

Section 4

ECONOMICS OF FEEDING STRAW

Chapter 4.1

FEEDING OF UREA TREATED STRAW IN THE TROPICS.
I. A REVIEW OF ITS TECHNICAL PRINCIPLES AND ECONOMICS[1]

J.B. Schiere and A.J. Nell

SUMMARY

Urea treatment is a method in which straw is treated by ammonia released from urea. The process is similar to ammonia treatment, and it is a technically feasible method to improve the nutritive value of straw. Application in the field depends on economic and practical considerations. Local prices of feed and produce as well as type and level of animal production determine whether there is any economic advantage in feeding treated (TS) over untreated straw (US). Two approaches are used to assess the economics of straw treatment: (a) comparison of the cost of a unit of energy (TDN) and crude protein (CP); (b) the use of least cost ration formulation (LCRF). The first approach is convenient but can be misleading because of its simplifications. The use of LCRF with linear programming can account for more factors, such as dry matter intake limitations. The calculations show that straw treatment is economically attractive (a) when treated straw is cheap compared with other supplements for cows of medium production, and (b) when animal products can be sold at a remunerative price. Secondary effects of treatment on health, calf rearing or composition of produce are reviewed and no negative effects are known. Aspects of ammonia economy and savings of concentrate as well as the use of straw for other purposes are discussed. The emphasis of this article is on urea treatment of rice straw for the tropical smallholder's farming system, especially south and southeast Asia, with reference to work from elsewhere.

---

[1] Published in Animal Feed Science and Technology, 43(1993): 135-147.

INTRODUCTION

The decreasing availability of communal grazing land and increased cropping intensity make livestock feeding in many systems increasingly dependent on crop residues such as straws. The potential of crop residues for animal feed is large in quantitative terms. With a grain/straw ratio of rice of around 1.3 (Kossila, 1984) and an average grain yield per harvest of approximately 3000 kg.ha^{-1} (De Geus, 1973), the straw yield of 1 ha provides sufficient dry matter (only in terms of quantity) for year-round feeding of at least one small tropical livestock unit of 350 kg liveweight. However, its concentration of digestible nutrients in straw is low and hence the dry matter intake alone is not a good measure of nutrient intake. The low nutrient content of straw limits its use for animal feed.

Practical options to overcome this problem of low nutritive value are reviewed in Sundstøl and Owen (1984), Owen and Jayasuriya (1989) and Kiran Singh and Schiere (1991). These practical options include:
-   supplementation with limiting nutrients;
-   chemical or physical treatment.

Other options are discussed by Berger *et al.* (1979), Zemmelink (1986), Capper (1988), Reed *et al.* (1988), Wahed *et al.* (1990) and Kiran Singh and Schiere (1991). They include:
-   allowing selective consumption;
-   better post/pre-harvest management of the straw;
-   upgrading of straw quality through plant breeding.

Only the economics of supplementation and/or treatment to overcome the problem of low nutritive value of straw will be discussed in this paper, even though an economic assessment of straw treatment should compare all other options, including the use of straw for non-feed purposes, such as the use of straw for roofing, fertiliser, or production of fuel, chemicals and mushrooms (Staniforth, 1982; White, 1984; Zadrazil, 1984; Hartley *et al.*, 1987; Rajarathnam and Zakia Bano, 1989).

This is the first article in a series of three. It briefly reviews the technical background of treatment *vs.* supplementation. Subsequently it compares the economics of feeding treated straw (TS) with that of untreated straw (US), based on a comparison of the cost of a unit of energy (TDN) and crude protein (CP), to be followed by the use of least cost ration formulation (LCRF). The technical validity of ration formulation by comparing calculated performance with actual experimental data from feeding trials, is discussed in part II (Schiere and De Wit, 1995). The application of ration formulation and feeding standards in farming systems is discussed in a third paper (Schiere and De Wit, 1993).

TREATMENT AND SUPPLEMENTATION

Several treatments are available to improve the nutritive value of straw. Chemical treatment with ammonia has replaced the more effective treatment by NaOH (Sundstøl and Owen, 1984; Wanapat *et al.*, 1985). An attractive source of ammonia in the tropics is urea (Jayasuriya and Perera, 1982; Perdok *et al.*, 1982; Schiere and Ibrahim, 1989), but in cooler climates urea is not as effective as ammonia (Westgaard and Sundstøl, 1986). Also, large farmers in the (sub)tropics may prefer ammonia when it is available, as is the case in Egypt (Barker *et al.*, 1987). The exact nature of the treatment with ammonia is not well understood (Neilson and Stone, 1987; Chesson, 1988) but the basics are described in Sundstøl and Owen (1984). It is not certain to what extent the improved nutritive value of treated straw is a result of the supply of NPN and to what extent it is a result of changes in the structure of the straw. The effect of urea supplementation also depends on the composition of the rest of the ration and the amount of urea added. The effect of treatment cannot be accounted for by NH_3 supplementation only (Schiere and Ibrahim, 1986; Baber *et al.*, 1988; Djajanegara and Doyle, 1989).

Supplementation of poor quality roughages is done by feeding limiting nutrients in the form of concentrates, special minerals, proteins or green forages. It aims at one or a combination of two distinct objectives.

- Feeding for a positive associative effect: this approach uses small quantities of supplements such as minerals or proteins to enhance rumen fermentation leading to increased intake and digestibility (Schiere and De Wit, 1993a). The primary objective is utilization of available roughage. It generally implies low levels of animal production, e.g. survival feeding which is essential and valuable in many tropical farming systems. This approach assumes a good availability of (cheap) roughages. Supplementation for positive associative effects is done for example when straw is not treated but only supplemented with urea (Perdok *et al.*, 1982; Schiere and Wieringa, 1988; Van Der Hoek *et al.*, 1989) or when urea molasses lickblocks are fed (Leng *et al.*, 1991). It will not be discussed further here because it is outside the scope of this paper.
- Substitutional supplementation, which aims to reach a desired level of animal production with moderate to high levels of supplement, often by substituting a part of the basal feed in the ration (Creek *et al.*, 1984; Ghebrehiwet *et al.*, 1988; Tharmaraj *et al.*, 1989). This supplementation can even be done at the expense of optimum biological processes in the rumen, as is the case with many tropical urban milk producers who feed high concentrate levels.

Effects of straw treatment on level and quality of animal production

The feeding of TS instead of US increases individual animal production, but the magnitude of the increase depends on factors such as the nutritive value of the other components of the ration, age and type of livestock, level and type of product and disease incidence (Ibrahim 1986a). There has been little systematic work carried out on the effect of TS on the quality and composition of animal produce and other side effects, and there is no information available on any negative side effects of feeding TS compared with US.

The feeding of TS increases milk yield per day, persistency of lactation and liveweight gains (Khan and Davis, 1981; Perdok *et al.*, 1982; Chemjong, 1991). Farmers and formal research have observed that the butterfat content of milk remains unaffected (Khalid, 1988; Icbal, 1989) or increases (Perdok *et al.*, 1982; Hermansen, 1983; Kristensen, 1984; Rai and Mudgal, 1988; Van Der Hoek *et al.*, 1989).

Liveweight gains on TS are around 100-150 g day^{-1} for cattle of 100-200 kg liveweight (Perdok *et al.*, 1982; Creek *et al.*, 1984; Ghebrehiwet *et al.*, 1988; Tharmaraj *et al.*, 1988). Little is known about the effect of feeding TS on dressing percentage, though there is no reason to expect a negative effect (Haque and Saadullah, 1983; Saadullah, 1986). The liveweight gain of calves from cows fed on TS is higher than that when cows are fed US (Perdok *et al.*, 1982; Van Der Hoek *et al.*, 1989). There are no indications that feeding TS has a negative effect on animal health and fertility (Perera, 1986; Sewalt and Schiere, 1989).

Economics of treatment or supplementation

The choice between treated straw with or without supplements and untreated with or without supplements is determined by financial and practical considerations. Straw treatment is a simple technique that can be applied in different ways according to local circumstances (Schiere and Ibrahim, 1989; Kiran Singh and Schiere, 1991). However, farmers can reject new technology for reasons of unknown risks, natural reluctance, unfavourable economics or because the technology does not fit in with their family labour availability.

An expensive and laborious way to determine economics is to compare rations in on-station feeding trials (Kaasschieter *et al.*, 1983; Creek *et al.*, 1984; Barker *et al.*, 1987), or with on-farm testing (Ibbotson *et al.*, 1984; Barker *et al.*, 1987). However, a major disadvantage of on-farm or on-station feeding trials is that they can never cover the entire range of rations for all specific field situations. Field application and extension for highly variable farming systems needs an approach that can predict economics of rations for different situations (Potts, 1982; Amir, 1986). The following pages summarise and compare the use and results of two such approaches, as developed by the Straw Utilisation Project of Sri Lanka. The calculations were done together with intensive on-farm monitoring of practical aspects of straw treatment. The details of the calculations are described in detail by Nell (1986) and Schiere and Ibrahim (1989), but are summarised here in Tables 1 and 2.

A simple method compares the unit cost of nutrients, as done by Kearl (1982), also being the basis of the calculations by Mallorie and Ali (1987). Differences between feeds and farming systems are shown in Table 1 where the rural farming system is located in the Sri Lankan hills and where the peri-urban farming system is located in Trivandrum in South India. The essential difference between the two systems is in the roughage/concentrate price ratio and not in the difference between countries. This method of cost per unit of feed value is simple but inadequate because it values the energy and protein separately, and it does not take into account the dry matter intake limitations and aspects such as substitution rates and associative effects.

A more comprehensive approach is the use of LCRF. By using linear programming it is relatively simple to incorporate DMI limitations, substitution rates, feed values, etc. The technical as well as socio-economic validity of LCRF for straw feeding is discussed further by Chesworth *et al.* (1989) and Schiere and De Wit (1993a,b) who found that ration formulation with the National Research Council (NRC) values tend to overestimate responses, but not the ranking. Our numerous calculations with linear programming on a large set of different values constituted a sensitivity analysis that gave a consistent and logical picture (Table 2). It showed that the feasibility of straw treatment and supplementation depends on the cost of feeds and on the level of production. The feeding of treated straw is most attractive in farming systems with:

- a low price ratio of (treated) straw over other feeds;
- cows of a reasonable milk production;
- a good milk market.

As a rule of thumb it can be said that the cost of treated straw should not be more than half the cost of concentrate on a dry matter basis. The exact ratio obviously depends on feeding values and animal production levels. Our calculations, based on animals of 350 kg liveweight, assume a level of milk production in many rural tropical areas between 5 and 10 l.animal^{-1}.day^{-1} (Cunningham and Syrstad, 1987), partly based on the smaller animal size. The results make sense. Even high production is not financially attractive when the milk cannot be sold at a remunerative price (Schiere and De Wit, 1993). The production levels from Table 2 are to be taken as indications, and not as absolute values, since they depend on price ratios and animal size in the prevailing farming system. The calculations are approximations and a small feed cost advantage of TS over US rations, as shown (especially for the peri-urban system) in Table 2, does not warrant the introduction of a (new) technology. The larger feed cost differences for rural situations, with 8 l day^{-1} milk production, indicate a cost difference that makes introduction of a new technology attractive.

Table 1.   Average costs of nutrients (US$ per 100 kg) in two different farming systems (based on Nell, 1986)

	Rural farming system			Peri-urban farming system		
	DM	TDN	CP	DM	TDN	CP
US	0.21	0.54	5.21	6.66	17.54	166.67
TS	1.67	3.71	16.67	8.33	18.50	83.33
Grass	2.08	3.79	20.83	8.33	15.16	83.33
Rice bran[1]	2.50	6.25	41.67	6.25	15.62	104.16
Coconut cake	10.42	14.88	52.08	20.83	29.75	104.16
Commercial conc.	14.58	22.42	97.21	14.55	22.19	97.08

[1] In the rural farming system, the rice bran is of very poor quality (Ibrahim, 1987)

Table 2.   Cost of rations (US$ day^{-1} per animal) based on untreated straw (US) and treated straw (TS) in two different farming systems (based on Nell, 1986).

	Rural farming system				Peri-urban farming system			
	Cheapest ration	Cost day^{-1}	Most expensive ration	Cost day^{-1}	Cheapest ration	Cost day^{-1}	Most expensive ration	Cost day^{-1}
Maintenance	US+	0.02	TS-	0.09	US+	0.41	TS+	0.43
Milk (4 l) + maintenance	US+	0.14	TS-	0.16	TS+	0.63	US+	0.69
Milk (8 l) + maintenance	TS+	0.21	US+	0.32	TS+	0.86	US+	0.98

+/-, with or without supplement.
The milk price is $0.15 l^{-1} in the rural farming system and $0.25 l^{-1} in the peri-urban farming system.

The results in Table 2 can be partly explained by looking at the unit cost of nutrients in Table 1. Since maintenance can be obtained from US with little supplement, the small additional cost of the supplement is likely to be less than the cost of treatment. At higher levels of production, the US ration requires progressively more and better supplements than the TS ration. If straw is cheap and supplements are expensive, the higher intake of TS saves supplement by substituting it with straw. This results, of course, only in savings at a low treated straw/concentrate cost ratio. In a rural system a TS ration is attractive because of savings on relatively expensive supplements at medium production levels. In the peri-urban system the difference in cost between straw and supplements is less and feeding of TS is also less attractive. Moreover, for the higher levels of production that are common in peri-urban systems, even the intake of TS is not high enough to provide sufficient nutrients. At those levels, the animals need to be fed increasing amounts of concentrates and the difference between US and TS rations decreases. Roughage feeding is then done only to satisfy minimum fibre requirements. The proportion of supplements on the ration is high because they are cheap sources of nutrients and because more can be fed before maximum DMI is reached. If only small amounts of straw are included in the ration at higher levels of production the intake and quality differences in terms of digestible energy and protein are small between TS and US rations.

In situations of extreme feed shortage where the sole aim of feeding is survival, the treatment of straw enables more nutrients to be obtained from the same batch of feed, even without utilising the extra intake. This situation arises during droughts and in rangeland conditions, such as for example in Australia (G.McL. Dryden, personal communication, 1988).

OTHER ASPECTS OF STRAW TREATMENT ECONOMY

Many other aspects can be included in the economic considerations. Many of them are not sufficiently substantiated or quantified, but they need to be discussed at some length here. They concern the savings of concentrate per unit urea, secondary effects of treatment, the use of urea for fertiliser *vs.* treatment, environmental aspects and the usefulness of TS for small *vs.* large farmers. Urea ammonia treatment can save on supplements, depending on the level of production of the animals and on the feeding value of the treated straw and concentrates. It is estimated that from 5 to 14 kg concentrate can be saved kg^{-1} NH$_3$, depending on the level of production (Creek *et al.*, 1984; Schiere, 1988). Such a saving is a combined effect of improved quality and intake of straw after treatment and is therefore economically beneficial only if enough straw is available. Vijayalaksmi *et al.* (1988) estimated that urea ammonia treatment saves 550 kg of concentrate per lactation per cow at the expense of 110 kg urea and 660 kg of extra straw intake when animals are producing 7 l milk day^{-1} on average over a lactation.

Secondary effects of treatment on cost of feeding can be positive and negative. A farmer can lose straw because of moulding after treatment, but moist straw can potentially be preserved by urea. The intake of TS is larger than that of US and many farmers find that wastage of (unchopped) TS is lower than that of US. It is possible that a small amount of alkaline TS buffers rumen pH better than US, an aspect that might play a role in high concentrate rations. Such effects are not included in the calculation mentioned above.

Negative secondary effects of feeding TS over US in regard to animal (re)production are unlikely as discussed earlier.

Alternative uses of urea, e.g. as a crop fertiliser, should also be taken into account when economics are considered. Nitrogen fertiliser responses are around 20 kg rice kg^{-1} N and 10-27 kg wheat kg^{-1} N (De Geus, 1973). Moreover, urea used as fertiliser for grain can produce more bran and more straw with a slightly higher feeding value (Ibrahim *et al.*, 1988). The calculations become more complicated when it is taken into consideration that the urea-N used for the treatment of straw is not necessarily all lost, since part of it can be recovered in dung and urine. Because only a part of the total ammonia used for treatment is chemically bound to the treated straw, important losses of N after treatment with urea occur upon feeding or aeration of the freshly treated straw (Sundstøl and Coxworth, 1984; Rai and Gupta, 1989). Jayasuriya and Perera (1982) show that only one-third of the total NH$_3$ applied for straw treatment is released in 2 hours under normal ambient temperature and ventilation. This environmentally and economically undesirable loss can technically be reduced by trapping the excess ammonia in water or acid, but these processes are not practical in tropical field conditions. Whether urea is more efficient as a fertiliser for grain and straw than as a treatment for straw, depends mainly on the relative prices of grain, straw, milk, concentrate and urea. The question of urea use for either fertiliser or straw treatment is only relevant when the availability of urea is limited, in which case, the use of urea for food crops will generally be favoured.

An alternative use of straw to that as an animal feed is its direct incorporation into the soil to improve soil structure. In both cases it is desirable to add N (Staniforth, 1982). Except for the effect of mulching, there seems to be little difference in soil fertility whether straw is used as animal feed or directly incorporated into the soil. If the dung collection and application is done properly, the losses of nutrients in both processes are of the same magnitude (Verschuur, 1991).

Energy return is positive if the use of fossil energy for the production of TS is compared with the production of other feeds, including energy costs for fencing and fertiliser in Western conditions (Sundstøl and Coxworth, 1984). Otherwise, with an energy requirement of 60 MJ for the production of 1 kg urea (Lockeretz, 1980), and considering that 1 kg of urea saves 5 kg of grain (energy requirement approximately 34.4 MJ kg^{-1}), the balance is in favour of urea for treatment. Interestingly, if treatment makes possible the use of straw as feed rather than having to burn or discard it, the efficiency of urea addition is more than 100%, since both N and captured solar energy are not lost in smoke.

Whether small farmers will benefit more from this technology than large farmers, or women and children more than men, is system specific. Some Sri Lankan women were happy with the use of treated straw as cattle feed because it saved labour, in cutting and carrying grass, though in their case the attraction of feeding TS was probably more in the introduction of straw as a hitherto unused feed than in the treatment itself. Later on, many stopped treatment of straw but continued the use of straw in time of scarcity. The use of urea, and the need to sell milk or meat to repay the cost, means that commercial farmers are likely to benefit most. The main factor is the availability of straw for the farmer in question. Small farmers in Iran and China are practising urea treatment (E.R. Ørskov and F. Dolberg, personal communication, 1991), especially when there is plenty of straw available. Large commercial farmers in Sri Lanka did not want to bother with the collection or purchase of large quantities of straw. They found the transport and labour cost too high. In urban dairies the proportion of straw in the ration is likely to be low, because of the relatively high transport costs for bulky materials. This explains why the expected benefits of straw treatment are low for both small and large urban dairy farmers. If straw is in abundant supply, but not commonly fed owing to its low quality as stipulated by Zemmelink (1986), then the increase in quality makes its use as feed more attractive and competition for its use will increase. In this process the small farmer generally loses, though it can be argued that improved possibilities of straw feeding will add value to the countryside.

CONCLUSIONS

The improvement of straw quality for ruminant feeding by urea treatment is effective and technically sound. The process is essentially simple but the applicability is limited because of economic or practical reasons. Animal production, in terms of growth and milk, is increased owing to the higher nutritive value and intake of straw after treatment. No negative secondary effects of urea treatment of rice straw are known. Whether the treatment of straw is economically more attractive than supplementation depends on the cost of feeds, the level of production and on practical considerations specific to the farming system. The use of LCRF shows that feeding of treated straw is most attractive for systems with cheap straw, expensive supplements, cows with reasonable production levels and good milk prices. Other aspects, such as the use of urea for crop production, environmental aspects and energetic efficiency are insufficiently elaborated but do not necessarily affect the economics in a negative sense. The use of urea as fertiliser for crops can be a more attractive proposition than its use on straw, an aspect that is particularly relevant when urea availability is restricted.

ACKNOWLEDGEMENTS

Thanks are due to J. De Wit for assisting in the preparation of the final manuscript, but particulary to the parties involved in the Straw Utilisation Project of Sri Lanka and the Indo-Dutch Bioconversion Project in India where most of this work was developed.

REFERENCES

Amir, P., 1986. Economical pre-screening of livestock technologies for small farmers in the nineties. Asian Livest., November: 147-150.

Baber, R.P., Matthewman, R.W. and Smith, A.J., 1988. Methods of increasing nutritional value of straw when used as food for goats, sheep, cattle and buffaloes. Dissertation Review III, Trop. Anim. Health Prod., 20: 99-102.

Barker, T.J., Yackout, H., Creek, M.J., Hathout, M. and El Nouby, H., 1987. Cattle feeding. Transfer of feeding systems from research phase to the farmer. World Anim. Rev., 61: 17-25.

Berger, L.L., Paterson, J.A., Klopfenstein, T.J. and Britton, R.A., 1979. Effect of harvest date and chemical treatment on feeding value of corn stalkage. J. Anim. Sci., 49: 1312-1316.

Capper, B.S., 1988. Genetic variation in the feeding value of cereal straw. Anim. Feed Sci. Technol., 21: 127-140.

Chemjong, P.B., 1991. Economic value of urea-treated straw fed to lactating buffaloes during the dry season in Nepal. Trop. Anim. Health Prod., 23: 147-154.

Chesson, A., 1988. Lignin-polysaccharide complexes of the plant cell wall and their effect on microbial degradation in the rumen. Anim. Feed Sci. Technol., 21: 219-228.

Chesworth, J.M., McKillop, D.F. and Spriggs, D.J., 1989. Combinations of agroindustrial by-products for use in dairy diets formulated by on farm use of a least cost ration system. In: A.N. Said and B.H. Dzowela (eds.), Overcoming Constraints to the Efficient Utilization of Agricultural By-products as Animal Feed. Proc. 4th Annual Workshop, 20-27 October 1987, Institute of Animal Research Mankon Station, Bamenda, Cameroun. ARNAB, Addis Ababa, pp. 61-79.

Creek, M.J., Barker, T.J. and Hargus, W.A., 1984. The development of a new technology in ancient land. World Anim. Rev., 51: 12-20.

Cunningham, E.P. and Syrstad, O., 1987. Crossbreeding *Bos indicus* and *Bos taurus* for milk production in the tropics. FAO Anim. Prod. Health Pap. 68, FAO, Rome, 90 pp.

De Geus, J.G., 1973. Fertilizer Guide for the Tropics and Subtropics. Centre de l'Etude de l'Azote, Zurich, 774 pp.

Djajanegara, A. and Doyle, P.T., 1989. Urea Supplementation compared with pretreatment. 1. Effects on intake, digestion and liveweight change by sheep fed on rice straw. Anim. Feed Sci. Technol., 27: 17-30.

Ghebrehiwet, T., Ibrahim, M.N.M., Schiere, J.B., 1988. Response of growing bulls to diets containing untreated or urea treated rice straw with rice bran supplementation. Biol. Wastes, 25(4): 269-280.

Haque, M. and Saadullah, M., 1983. Liveweight gain and gut contents of calves fed ammonia (urea)-treated or untreated rice straw with and without water hyacinth. In: T.R. Preston, C.H. Davis, M. Haque and M. Saadullah (eds.), 1983. Maximum livestock production from minimum land. Proc. 4th Seminar, 2-4 May 1983, Bangladesh Agricultural University, Mymensingh, pp. 45-46.

Hartley, B.S., F.R.S, Broda, P.M.A. and Senior, P.J., (eds.), 1987. Technology in the 1990's: Utilization of Lignocellulosic Wastes. Philos. Trans. R. Soc. London, 321: 403-568.

Hermansen, J.E., 1983. NH₃-behandlet contra ubehandlet halm i blandinger til malkekoer. Statens Husdyrbrugsforsog, Copenhagen. Medd. No. 451, 4 pp.

Ibbotson, C.F., Mansbridge, R. and Adamson, A.H., 1984. Commercial experience of treating straw with ammonia. Anim. Feed Sci. Technol., 10: 223-228.

Ibrahim, M.N.M, 1986. Rice bran as a supplement for straw based rations. In: R.M. Dixon (ed.), Ruminant Feeding Systems Utilizing Fibrous Agricultural Residues. Proc. 6th Ann. Workshop AFARR, 1-3 April 1986, Los Baños, Philippines. International Development Program of Australian Universities and Colleges, Canberra, A.C.T.

Ibrahim, M.N.M., 1987. Rice bran as a supplement for straw based rations. In: Dixon, R.M. (Ed.), Ruminant feeding systems utilizing agricultural residues-1986. Proceedings of the 6th annual workshop of the AAFARR Network held in the University of the Philippines at Los Banos, 1-3 April, 1986. IDP, Canberra, pp. 139-146.

Ibrahim, M.N.M., Tharmaraj, J. and Schiere, J.B., 1988. Effect of variety and nitrogen application on the nutritive value of rice straw and stubble. Biol. Wastes, 24: 267-274.

Iqbal, A., 1989. Effect of feeding urea-treated wheat straw on production and composition of milk in Nili-Ravi buffaloes at station (farm) level. M.Sc. Thesis, Faculty of Animal Husbandry, University of Agriculture, Faisalabad, Pakistan, 66 pp.

Jayasuriya, M.C.N. and Perera, H.G.D., 1982. Urea ammonia treatment of rice straw to improve its nutritive value for ruminants. Agric. Wastes, 4: 143-150.

Kaasschieter, G.A., Perdok, H.B., Van Wageningen, N.M. and Jayasuriya, M.C.N., 1983. Economic feasibility of (urea treated) straw based rations for lactating Surti buffaloes in Sri Lanka. In: E.H. Ketelaars and S. Boer Iwema (eds.), Animals as Waste Converters. Proc. Int. Symp. 30 November-2 December 1983, Wageningen. Pudoc, Wageningen, pp. 72-72.

Kearl, L.C., 1982. Nutrient requirements of ruminants in developing countries. International Feedstuff Institute, Utah State Institute, Logan, 381 pp.

Khalid, F.A., 1988. Effect of feeding urea-treated wheat straw on milk production and milk composition. M.Sc. Thesis, Faculty of Animal Husbandry, University of Agriculture, Faisalabad, Pakistan, 45 pp.

Khan, A.K.M.N. and Davis, C.H., 1981. Effect of treating paddy straw with ammonia (generated from urea) on the performance of local and crossbred lactating cows. In: M.G. Jackson, F. Dolberg, C.H. Davis, M. Haque and M. Saadullah (eds.), Maximum Livestock Production from Minimum Land. Proc. Seminar, 2-5 February 1981, Bangladesh, Mymensingh Agricultural University, Bangladesh, pp. 168-180.

Kiran Singh and Schiere, J.B. (eds.), 1991. Feeding of ruminants on fibrous crop residues. Proc. Int. Workshop, 4-8 February 1991, NDRI Karnal. ICAR, New Delhi, India, in press.

Kossila, V.L., 1984. Location and potential feed use. In: F. Sundstøl and E. Owen (eds.), Straw and Other Fibrous By-products as Feed. Developments in Animal and Veterinary Sciences, 14. Elsevier, Amsterdam, pp. 4-24.

Kristensen, V.F., 1984. Straw etc. in practical rations. In: F. Sunstol and E. Owen (eds.), Straw and Other Fibrous By-products as Feed. Developments in Animal and Veterinary Sciences, 14. Elsevier, Amsterdam, pp. 431-453.

Leng, R.A., Preston, T.R., Sansoucy, R. and Kunju, P.J.G., 1991. Multinutrient blocks as a strategic supplement for ruminants. World Anim. Rev., 67(2): 11-20.

Lockeretz, W., 1980. Energy inputs for nitrogen, phosphorous and potash fertilizers. In: D. Pimentel (ed.), Handbook of Energy Utilization in Agriculture. CRC Press, Boca Raton, FL, p. 24.

Mallorie, E.R. and Ali, A., 1987. Viability of systems to exploit non-conventional feed resources by small farmers. In: P. Amir, A.S. Akhtar and M.D. Dawson (eds.), Livestock in Pakistan Farming System Research. Proc. Workshop, 8-15 April 1987, Islamabad, Agricultural Research Council, Islamabad, Pakistan, pp. 47-59.

Neilson, M.J. and Stone, B.J., 1987. Chemical composition of fibrous feeds for ruminants in relation to digestibility. In: T.F. Reardon *et al.* (eds.), Proc. 4th AAAP Animal Science Congress, Hamilton, New Zealand, pp. 63-66.

Nell, A.J., 1986. Feed cost in different farming systems. In: M.N.M. Ibrahim and J.B. Schiere (eds.), Rice Straw and Related Feeds in Ruminant Rations. Proc. Int. Workshop, 24-28 March 1986, Straw Utilization Project, Kandy, Sri Lanka/Agricultural University Wageningen, Netherlands, pp. 326-336.

Owen, E. and Jayasuriya, M.C.N., 1989. Use of crop residues as animal feeds in developing countries. Res. Dev. Agric., 6(3): 129-138.

Perdok, H.B., Thamotharam, M., Blom, J.J., Van Den Born, H. and Van Veluw, C., 1982. Practical experiences with urea-ensiled straw in Sri Lanka. In: T.R. Preston, C.H. Davis, M. Haque and M. Saadullah (eds.), Maximum Livestock Production from Minimum Land. Proc. 3rd Seminar, 13-18 February 1982, Bangladesh Agricultural Research Institute, Joydebpur, Bangladesh, pp. 123-134.

Perera, B.M.A.O., 1986. Health aspects of feeding (un)treated straw. In: M.N.M. Ibrahim and J.B. Schiere (eds.), Rice Straw and Related Feeds in Ruminant Rations. Proc. Int. Workshop, 24-28 March 1986, Straw Utilization Project, Kandy, Sri Lanka/Agricultural University Wageningen, Netherlands, pp. 350-355.

Potts, G.R., 1982. Application of research results on by-product utilization: economic aspects to be considered. In: Berhane Kiflewahed, G.R. Potts and R.M. Drysdale (eds.), By-product Utilization for Animal Production. Proc. Workshop on Applied Research, 26-30 September 1982, Nairobi, Kenya. IDRC, Ottawa, Ont., pp. 116-127.

Rai, S.N. and Gupta, B.N., 1989. Effect of long term feeding of urea (ammonia) treated wheat straw on intake, nutrient utilization, N metabolism and economics of feed cost in yearling murrah buffalo bulls. Buffalo J., 4: 49-62

Rai, S.N. and Mudgal, V.D., 1988. Feeding treated straw for efficient utilization and production by Murrah buffaloes. Buffalo J., 2: 161-172.

Rajarathnam, S. and Zakia Bano, 1989. *Pleurotus* mushrooms. Part III. Biotransformations of natural lignocellulosic wastes: commercial applications and implications. Crit. Rev. Food Sci. Nutr., 28: 31-113.

Reed, J.D., Capper, B.S. and Neate, P.J.H. (eds.), 1988. Plant Breeding and the Nutritive Value of Crop Residues. Proc. Workshop, 7-10 December 1987, ILCA, Addis Ababa, Ethiopia, 334 pp.

Saadullah, M., 1986. Supplementing urea treated rice straw with fish meal and its effects on the gut contents and dressed carcass weight. In: M.N.M. Ibrahim and J.B. Schiere (eds.), Rice Straw and Related Feeds in Ruminant Rations. Proc. Int. Workshop, 24-28 March 1986, Straw Utilization Project, Kandy, Sri Lanka/Agricultural University Wageningen, Netherlands, pp. 248-254.

Schiere, J.B., 1988. Crop by-products, their potential and limitations especially for ruminant feed. In: M. Soejono, A. Musofie, R. Utotmo, N.K. Wardhani and J.B. Schiere (eds.), Limbah Pertanian Sebagai Pakan dan Manfaat Lainnya (Crop Residues for Feed and Other Purposes). Proc. 2nd Workshop, 16-17 November 1987, Bioconversion Project, Balai Penelitian Ternak, Grati, Indonesia, pp. 10-20.

Schiere, J.B. and De Wit, J., 1993. Feeding standards and feeding systems, Anim. Feed Sci. Technol., 43: 121-134.

Schiere, J.B. and De Wit, J., 1995. Feeding of urea ammonia treated straw in the tropics. Part II: Assumption on nutritive values and their validity for least cost ration formulation. Anim. Feed Sci. Technol., 51:45-63.

Schiere, J.B. and Ibrahim, M.N.M., 1986. Supplementation of straw rations, a practical approach. In: M.N.M. Ibrahim and J.B. Schiere (eds.), Rice Straw and Related Feeds in Ruminant Rations. Proc. Int. Workshop, 24-28 March 1986, Straw Utilization Project, Kandy, Sri Lanka/Agricultural University Wageningen, Netherlands, pp. 275-281.

Schiere, J.B. and Ibrahim, M.N.M., 1989. Feeding of urea-ammonia treated rice straw. A compilation of miscellaneous reports produced by the Straw Utilization Project (Sri Lanka). PUDOC, Wageningen, Netherlands, 125 pp.

Schiere, J.B. and Wieringa, J., 1988. Overcoming the nutritional limitations of rice straw for ruminants. 2. Response of growing Sahiwal and local cross heifers to urea upgraded and urea supplemented straw. Asian-Austr. J. Anim. Sci., 1(4): 209-212.

Sewalt, V.J.H. and Schiere, J.B., 1989. Health problems associated with feeding urea treated rice straw. Paper presented at the 6th Conf. on Tropical Animal Health and Production, 28 August-1 September 1989, Wageningen. University of Utrecht, Utrecht.

Staniforth, A.R., 1982. Straw for Fuel, Feed and Fertilizer? Farming Press, Ipswich, UK, 153 pp.

Sundstøl, F. and Coxworth, E.M., 1984. Ammonia treatment. In: F. Sundstøl and E. Owen (eds.), Straw and Other Fibrous By-products as Feed. Developments in Animal and Veterinary Sciences, 14. Elsevier, Amsterdam, pp. 196-247.

Sundstøl, F. and Owen, E. (eds.), 1984. Straw and Other Fibrous By-products as Feed. Developments in Animal and Veterinary Sciences, 14. Elsevier, Amsterdam, 604 pp.

Tharmaraj, J., Van Der Hoek, R., Sewalt, V.J.H. and Schiere, J.B., 1989. Overcoming the nutritional limitations of rice straw for ruminants. 4. Urea ammonia treatment and supplementation with *Gliricidia maculata* for growing Sahiwal bulls. Asian-Austr. J. Anim. Sci., (2): 85-90.

Van Der Hoek, R., Muttetuwegama, G.S. and Schiere, J.B., 1989. Overcoming the nutritional limitations of rice straw for ruminants. 1. Urea ammonia treatment and supplementation of straw for lactating Surti buffaloes. Asian-Austr. J. Anim. Sci., 1(4): 201-208.

Verschuur, C.M., 1991. Composting *versus* digestion. A study to the effect of composting compared with digestion by ruminants on the composition of biomass residue. M.Sc. Thesis, Department of Tropical Animal Production, Agricultural University Wageningen, Netherlands, 28 pp. (in Dutch).

Vijayalaksmi, S., Amirth Kumar, M.N. and Bhaskar, B.V., 1988. Target group identification by using least cost ration formulation. In: Kiran Singh and J.B. Schiere (eds.), Fibrous Crop Residues as Animal Feed. Proc. Int. Workshop, 27-28 October 1988, NDRI Bangalore. ICAR, New Delhi, India, pp. 231-238.

Wahed, R.A., Owen, E., Neate, M. and Hosking, B.J., 1990. Feeding straw to small ruminants: effect of amount offered on intake and selection of barley straw by goats and sheep. Anim. Prod., 51: 283-289.

Wanapat, M., Sundstøl, F. and Garmo, T.H., 1985. A comparison of alkali treatment methods to improve the nutritive value of straw. I. Digestibility and metabolizability. Anim. Feed Sci. Technol., 12: 295-309.

Westgaard, P. and Sundstøl, F., 1986. History of straw treatment in Europe. In: M.N.M. Ibrahim and J.B. Schiere (eds.), Rice Straw and Related Feeds in Ruminant Rations. Proc. Workshop, 24-28 March 1986. Straw Utilization Project, Kandy, Sri Lanka/Agricultural University, Wageningen, Netherlands, pp. 181-191.

White, D.J. (ed.), 1984. Straw Disposal and Utilization. A Review of Knowledge. MAFF, London, 94 pp.

Zadrazil, F., 1984. Microbial conversion of lignocellulose into feed. In: F. Sunstol and E. Owen (eds.), Straw and Other Fibrous By-products as Feed. Developments in Animal and Veterinary Sciences, 14. Elsevier, Amsterdam, pp. 276-292.

Zemmelink, G., 1986. Measuring intake of tropical forages. In: C.C. Balch and A.J.H. Van Es (eds.), Recent Advances in Feed Evaluation and Rationing Systems for Dairy Cattle in Extensive and Intensive Production Systems. International Dairy Federation, Brussels, pp. 17-21.

FEEDING UREA AMMONIA TREATED RICE STRAW IN THE TROPICS. II: ASSUMPTIONS ON NUTRITIVE VALUE AND THEIR VALIDITY FOR LEAST COST RATION FORMULATION[1]

J.B. Schiere and J. De Wit

SUMMARY

Field application of new feeding methods, such as urea ammonia treatment of straws requires an *ex-ante* assessment of their economic feasibility. Animal experiments are expensive for such an analysis, therefore cheaper and quicker methods are required. One such method is least cost ration formulation (LCRF), which extrapolates from a restricted number of feeding trials to a large number of feed combinations. The assumptions underlying LCRF can, however, be challenged and this paper discusses their validity in nutritional terms. The first part reviews literature on nutritive quality of rice straw, in relation to nutrient requirements and animal performance on rice straw based diets. The second part calculates the expected performances (P_{calc}) on straw based rations, using measured dry matter intake (DMI) and estimated total digestible nutrients (TDN) and crude protein (CP) values. It then compares P_{calc} with real performance (P_{real}) as found in the literature. Most dose response trials indicate a rather linear response to supplementation, with exceptions particularly at low levels of supplementation. The pooled values of P_{calc} relate linearly with P_{real} in several feeding trials. The P_{calc} only comes close to P_{real} when calculated on a TDN basis, when energy requirements for liveweight gain of light tropical animals are taken to be higher than those of NRC and when digestibility of energy in straws is assumed to be low. This suggests a low utilisation of digestible energy from either treated or untreated straw for reasons that cannot easily be established. Total CP supply cannot be limiting because the P_{calc} on the basis of CP also considerably overestimates P_{real}. The available literature values do not allow specification of CP into fractions according to rumen degradability. At production levels around maintenance, the correlation between P_{real} and P_{calc} decreases, possibly because at those levels the variability of maintenance requirements exerts a bigger effect on the total performance. Also, at those levels, straw constitutes a larger part of the ration, and associative effects are more likely. The sparse literature values for milk production also indicate differences between P_{real} and P_{calc}. The conclusions are that, in the absence of better methods, LCRF can be useful for ranking purposes but absolute values need to be interpreted with care, and P_{calc} tends to overestimate P_{real}. Future performance trials should measure digestible organic matter intake, substitution rates and maintenance requirements by dose response curves in order to improve their predictive value, and to allow interpretation of differences between P_{real} and P_{calc}. Differences between P_{real} and P_{calc} can then be investigated and weighed against accuracy and cost of performance trials.

---

[1] Published in Animal Feed Science and Technology, 51 (1995): 45-63

INTRODUCTION

Animal nutrition research and extension require methods to assess the feasibility of technical innovations before their application under farm conditions. The most accurate method for the evaluation of new rations is the use of animal performance trials. However, the variability in the results of such trials that are often carried out with insufficient animals, the expense and time involved, besides the myriad of situations and rations that need to be tested, make it necessary to use simpler and quicker procedures (Schiere and De Wit, 1993). One such approach is the use of least cost ration formulation (LCRF) that aims to predict animal response from data on feed quality and animal requirements. If reliable, these procedures can be used for *ex-ante* analysis of economic feasibility, or for identification of farming systems where an innovation can properly be introduced (Schiere and Nell, 1993).

This paper is the second of a series of three that reports work to assess feasibility of the use of urea treated straw (TS) compared with the supplementation of untreated straw (US) in a livestock development project in Sri Lanka. The procedures for the treatment are explained in Schiere and Ibrahim (1989) and the need for this economic analysis arose from the pressure of donors, farmers and politicians to introduce quickly the use of TS. The sequence was therefore that LCRF was calculated on the basis of realistic but estimated values (Schiere and Nell, 1993). Afterwards, those results were tested with experiments as reported here. The validity of LCRF in socio-economic terms is discussed in a third paper (Schiere and De Wit, 1993).

The first part of this paper reviews literature on technical parameters for LCRF and animal responses. This provides a basis for the second part where expected animal performance (P_{calc}) is calculated and compared with real performance (P_{real}) from a number of growth and lactation trials with rice straw based rations. The calculations are based on CP and TDN, with the assumption that mineral and vitamin content is sufficient to meet animal requirements.

FEED QUALITY, ANIMAL REQUIREMENTS AND LCRF

Nutritional issues that affect LCRF with poor quality roughages, include uncertainty about:
- selective consumption of heterogenous feeds;
- dry matter intake (DMI) and substitution rates (SRs);
- variability of nutritive value within feed classes;
- requirements of protein and energy for rumen microbial growth, animal maintenance and production;
- associative effects.

Each of these issues is discussed below with emphasis on the use of treated and untreated rice straw in practical rations with varying levels of supplements.

Selective consumption of heterogeneous feeds

Selective consumption complicates estimation of quantity and quality of intake, for stall feeding as well as on pasture (Zemmelink, 1980; Roth *et al.*, 1990; Wahed *et al.*, 1990).

Selective consumption affects intake and digestibility of ingested feed, reported for straw eaten by goats (Wahed *et al.*, 1990) and sheep (Bhargava *et al.*, 1988). Digestibility in large cattle appears to be less affected by selective consumption in the case of treated compared with untreated rice straw (Badurdeen *et al.*, 1994). Some effect is, however, reported for large ruminants, particularly in the case of coarse untreated straws (Powell, 1985; Ulhas Prabhu *et al.*, 1988). Recent results with sheep fed on rice straw (Chuzaemi *et al.*, 1994) show an increase of organic matter (OM) digestibility at higher levels of excess feed.

Quality differences between selected plant parts are important but variable. Often leaves are more digestible than the stem, especially with coarse straws, but differences are sometimes nil or reverse. For example, some rice varieties have stems which are more digestible than the leaves (Capper, 1988; Walli *et al.*, 1988). Animals' preferences for less digestible leaves of rice straw (Chuzaemi *et al.*, 1992) could explain the decrease in digestibility at increased DMI levels in the work of Badurdeen *et al.* (1994). For the calculation of animal performance of large cattle on rice straw based rations in the second part of this paper we assume that digestibility is not affected by increased levels of feed offered. If this affects the calculations at all, it favours the response to US more than to TS (Badurdeen *et al.*, 1994), i.e. the positive response to TS is not overestimated.

Dry matter intake and substitution rates

Results of DMI trials with (treated) rice straws as presented in Table 1 are comparable with those in the review of Doyle *et al.* (1986, p. 103-106) but are high compared with those reported by the Agricultural Research Council (ARC, 1980), which predicts a DMI of 66.7 g kg$^{-0.75}$ for a coarse diet with a TDN value of 48. An approximate 25% increase in DMI due to urea treatment is common but variable, sometimes even between experimental periods of the same trial (Tharmaraj *et al.*, 1989). It is further complicated by supplementation effects and differences in the initial straw quality (Djajanegara and Doyle, 1989; Doyle and Panday, 1990). Generally, straw DMI decreases with supplementation while total DMI increases, but at low levels of supplementation with limiting nutrients the DMI of straw can increase (Doyle *et al.*, 1986, p. 102; Leng, 1990), a so-called associative effect.

Associative effects of supplementation are hard to predict. They can be positive as well as negative as discussed later. Positive associative responses depend on the type of supplementation and basal ration. They are theoretically less likely with high than with low quality roughages, as supported by the formula proposed by Tolkamp and Ketelaars (1992) in Figure 1. At higher levels of supplementation, substitution rates (SR) depend on the quality of the supplement and the basal feed, better basal feeds showing higher SR than poorer ones (ARC, 1980; Faverdin *et al.*, 1991). This trend, however, is not clear from the limited and variable data summarised in Table 2, probably because the SR there is calculated as a linear effect, possibly masking curvilinearity and initially positive responses.

More information about intake and SRs is essential for LCRF. Recent models for prediction of intake were formulated by the National Research Council (NRC, 1987) and Forbes (1988). Under average circumstances the differences between those models are relatively small but a model of Tolkamp and Ketelaars (1992) also gives predictions for extreme situations (Figure 1). The formula is derived from experiments with sheep and

Table 1   DMI values of US and TS from several experiments

DMI US (g kg$^{-0.75}$)	DMI TS (g kg$^{-0.75}$)	DMI TS (DMI US = 100%)	
85.0	110.5	130	Barton, 1987
79.2	101.5	128	Schiere et al., 1989[a]
75.7	97.9	129	Seanger et al.,1983
87.7	116.1	132	Tharmaraj et al. 1989, first period
79.0	80.6	102	Tharmaraj et al. 1989, second period[a]
77.1	99.1	129	Ghebrehiwet et al., 1988
89.6	115.9	129	Creek et al., 1984
103.8	112.1	108	Perdok, 1987[b]
56.5	71.0	126	Rai and Gupta, 1989
60.1	86.5	144	De Rond, unpublished
79.4	99.1	125	Average

[a] Average of two periods that were exactly the same.
[b] The small difference is possibly due to the high DMI of the US ration, which also contained substantial amounts of other, good quality feeds.

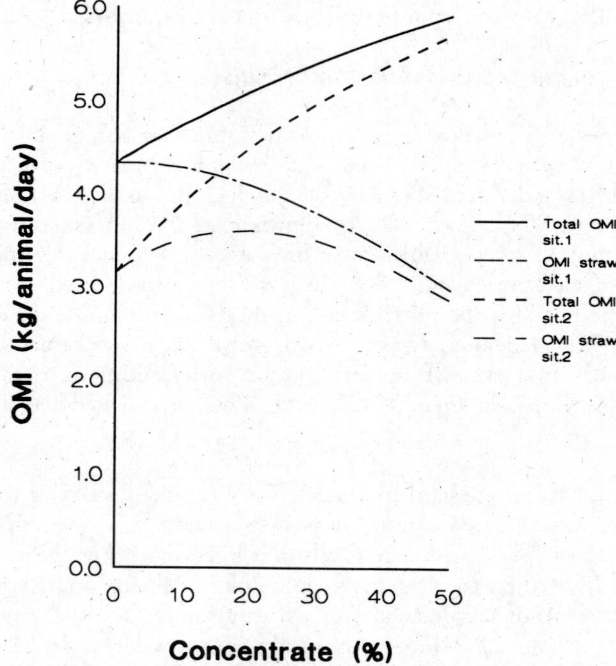

Figure 1. OMI (total and straw) with increasing percentage of supplement in the ration. Assumptions: DMI (g kg$^{-0.75}$) = F_a x F_p x (-19.5 + 0.05979 x CP_{om} + 92.46 x q) x (1.05/TDN) + F_c. Source: Tolkamp and Ketelaars (1992). F_a is a conversion factor from sheep to cattle; assumed 1.3 (ARC, 1980). F_p is a production factor, ranging from 1 to 2 according to level of milk production (ARC, 1980), here assumed to be 1; CP_{om} is grams CP/kilograms organic matter; q is metabolisability of the feed, here assumed to be TDN/1.2148. F_c is a correction factor concentrate: 0.37 g increase per kg$^{0.75}$ per percentage of concentrate (ARC, 1980). Situation 1 refers to an animal with a LW of 350 kg, straw: TDN 0.45, CP 40, OM 85%; concentrate: TDN 0.80, CP 120, OM 95%. Situation 2 is the same as Situation 1 but the TDN of straw is 0.35.

Table 2 Substitution rates (SR) in different experiments[a].

SR with US (kg kg⁻¹ supplement)	SR with TS (kg kg⁻¹ supplement)	
-0.226	NA	Schiere *et al.*, 1989
-0.52	-0.7	Tharmaraj *et al.*, 1989, first period
-0.39	-0.19[b]	Tharmaraj *et al.*, 1989, second period
-0.31	-0.25	Ghebrehiwet *et al.*, 1988
-0.25	-0.36	Creek *et al.*, 1984

[a] Average values assuming linear SR with increased DMI.
[b] At low DMI.

Table 3 TDN values of rice straw based on some in vivo experiments.

US	TS	
47.5	55.1	Schiere *et al.*, 1989[a]
51.2	55.6	Tharmaraj *et al.*, 1989[a]
45.4	50.5	Ghebrehiwet *et al.*, 1988[a]
51.2	56.4	Badurdeen *et al.*, 1986[b]
45.5	50.0	Soebarinoto *et al.*, 1990[c]

[a] Estimated from *in vivo* DMD measurements.
[b] Actual data based on OMD measurement with cattle.
[c] DOM measured with sheep.

requires two correction factors for our purposes (B.J. Tolkamp, personal communication, 1994). One is the animal factor (F_a), used for the conversion of the results from sheep to bovines, here assumed to be 1.3 based on ARC (1980). The second is the production factor (F_p). It adjusts the formula from growing animals to lactating animals. The F_p is assumed to be 1 for animals at maintenance and 2 for animals at peak lactation, values are again based on ARC (1980). The prediction of intake by this corrected formula also tallies with the actual intakes as reviewed in Table 1. Actual rather than predicted intake values are used in the calculations of the second part of this article for the comparison of P_{calc} and P_{real}, because the paper tests LCRF and not the intake formula.

Variability of nutritive value within feed classes

A large variation exists in digestibility values and other nutritional parameters, between and within cultivars for rice straw. The variation can be due to climate, cultivar, cutting height or stage of maturing, and it complicates the prediction of animal response (Doyle *et al.*, 1986, p. 102; Walli *et al.*, 1988; Soebarinoto *et al.*, 1990).

Most digestibility experiments with treated straw report only *in vitro* values, but some actual and estimated *in vivo* values are given in Table 3. The estimates appear to be high, considering the low energy availability in straw as found later in this article. They agree, however, with *in vivo* values estimated with acid insoluble ash (AIA) by Badurdeen *et al.* (1994) and Navaratne *et al.* (1990). A large between-animal variability of digestibility is common for low quality feeds (Van Soest, 1982; Doyle and Pearce, 1985; Cottyn *et al.*, 1989).

The problem of variable feeding values can be circumvented in economic calculations by applying a sensitivity analysis (Schiere and Ibrahim, 1989, Chapter 5). They used TDN values of 37% for US and 45% for TS in their calculations, based on determination of *in vivo* digestibility by total collection, using a 3 x 3 latin square with three replicates of young bulls on digestion crates for the evaluation of US, TS and straw supplemented with 2% urea (H.G.D. Perera, unpublished data). These low TDN values are used to predict animal production in this paper because they were the basis of the feasibility study as reported in Schiere and Nell (1993). They used low values to avoid a too optimistic estimate of animal performance towards the farming community. Even such low values will be shown to overestimate the energy utilisation for liveweight gain.

Average CP values of US range around 4% and for TS around 7%. The latter is often determined after oven drying and in practice more N is available to the rumen when TS is fed fresh. Jayasuriya and Perera (1982) report 10-12% CP in fresh TS 2 hours after presenting it to the animals. Crude protein in TS mainly consists of NPN, while a considerable part of the added ammonia N is so tightly bound to the straw that it can be considered to be unavailable to the rumen microbes (Sundstøl and Coxworth, 1984; Hvelplund, 1989). The variation in N-content and N-degradability of five different varieties of rice straw is presented in Table 4.

Table 4 Variation in N-content, solubility and digestibility of N, of five different varieties of rice straw, both treated and untreated (from Ibrahim *et al.*, 1989).

	N-content (g kg⁻¹ DM)	Water soluble N (%)	N-loss after incubation in the rumen (%)		
			48 h	72 h	240 h
US	5.8 - 13.1	28.2 - 36.3	38.0 - 48.9	38.5 - 54.5	44.2 - 63.5
TS	8.8 - 17.8	43.2 - 65.4	54.4 - 72.5	58.4 - 76.0	61.1 - 82.5

Requirements for rumen microbial growth, animal maintenance and production

The CP/TDN ratio in TS varies between 0.14 and 0.22 which is generally higher than the requirements for maintenance (0.15) and growth of older animals (0.13), but lower than the requirements for growth of young animals (0.27) and for milk production (0.27) (NRC, 1976). The CP/TDN ratio of US varies between 0.09 and 0.11, i.e. too low for all functions, even for maintenance.

The required CP/TDN ratio for microbial growth in the rumen is between 0.14 and 0.21, depending on the rumen degradability of CP, the rumen degradability of the OM, the recycling of N and the percentage of degradable protein which can be captured effectively in the rumen (Tamminga and Van Hellemond, 1977; Rohr *et al.*, 1979; ARC, 1980; Durand, 1989). There are indications that energy utilisation in the rumen is low for straw, i.e. that the efficiency of microbial nitrogen production is low, causing a relative protein shortage in the small intestine (Zorilla-Rios *et al.*, 1989; Doyle and Panday, 1990; Van Bruchem *et al.*, 1992). Van Der Hoek *et al.* (1989) showed that lactating buffaloes responded better to energy and protein supply than to protein supply alone. Most N for rumen function can be supplied by NPN sources (Kellaway and Leibholz, 1983). Small additions of true protein can be useful to provide a more steady supply of N or branched-chain carbon skeletons to the rumen fermentation (Durand, 1989; Leng, 1990; Oosting 1993).

A considerable effect on the nutrient requirements of animals is exerted by initial body condition, mature body weight, composition of growth, combined with environmental conditions, disease, starvation or breed differences (Elliot *et al.*, 1966; Frisch and Vercoe, 1978; Fox *et al.*, 1988; Laurenz *et al.*, 1991). The requirements for liveweight gain (LWG) of animals at different mature body weights need to be corrected, by using values for animals at a similar relative maturity, e.g. to correct for composition of liveweight gain (Taylor, 1985; Ogink, 1993). Such a correction is appropriate - albeit guesswork - for the experimental animals in many of the LWG trials reported in the literature. Requirements for milk production are commonly corrected according to the butterfat content.

The calculations in the second part of this paper, which compare P_{real} with P_{calc} are, for lack of better data, done with TDN and CP requirements per kilogram growth for 300 kg LW (3.8 kg TDN and 0.54 kg CP kg^{-1} growth) for animals of about 100 kg, with high estimates for maintenance (0.037 kg TDN kg$^{-0.75}$; 0.0056 kg CP kg$^{-0.75}$) (NRC, 1976). The only exception is the trial of Saadullah, where a requirement of 2.0 kg TDN and 0.42 kg CP kg^{-1} growth is assumed, owing to the young age and low weight of the animals in the trial. Requirements for milk production are taken to be 0.326 kg TDN and 0.087 kg CP kg^{-1} milk with 4% fat (NRC, 1978). All these requirements are slightly higher than the more recent estimates by NRC (1984, 1988).

Associative effects

Associative effects of supplementation on digestibility, intake, and utilisation of poor quality feeds are often reported (Coombe and Tribe, 1962; Schneider and Flatt, 1972; Mould *et al.*, 1983; Chenost and Reiniger, 1989; Leng, 1990). They imply a change in utilisation of one feedstuff by addition of another: a supplement therefore can have a larger or smaller effect on feed utilisation than could be expected on the basis of its nutrient content alone. The causes are many and not always well understood, hence the effects are even more difficult to predict (Cronjé, 1990; Leng, 1990). Apart from the action of nutritional or anti-nutritional factors, the associative effects are caused by the correction or creation of imbalances in rumen function or animal metabolism.

Associative effects can be positive with the supply of limiting nutrients, or negative with the feeding of excessive carbohydrates or anti-nutritional factors, causing a non-linear response to supplementation. This complicates LCRF, which often though not by definition, assumes linear additivity of nutrients. The aggregation of nutrients into CP and TDN as done in the calculations of this paper is likely to miss such responses to specific nutrients. The results from the dose response trials in Figures 2a-2i, and the fact that positive as well as negative associative effects tend to cancel each other, lead us to assume additivity. It is convenient for calculation purposes but likely to oversimplify the issue.

Associative effects are different for low levels of nutrition than for high levels. Positive effects might be expected to occur particularly on low quality rations, e.g. where the addition of energy, N or minerals can cause a substantial increase in nutrient utilisation (Perdok, 1987; Van Der Hoek *et al.*, 1989). This is, however, not clear from the response curves in Figures 2a-2i, which were mostly based on supplementation trials with rice bran and expeller cakes. Clear cases where associative effects are more positive with medium than with very poor quality basal feeds are reported by Doyle and Panday (1990).

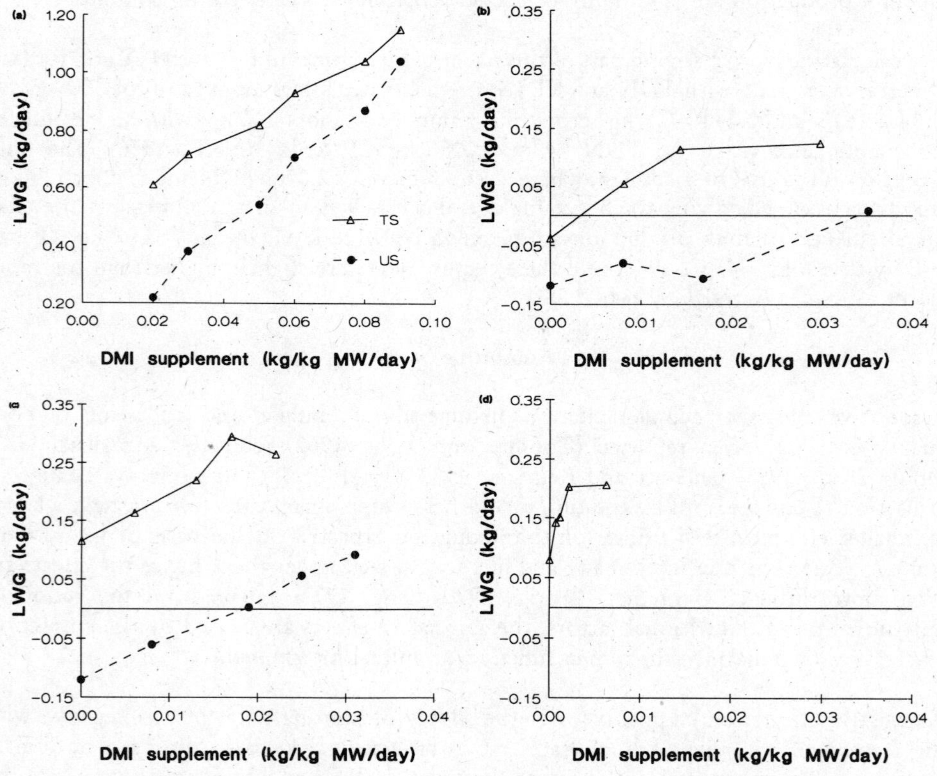

Figure 2. Review of dose response experiments on the relation between supplementation and liveweight gain (LWG). Sources: (a) Creek *et al.* (1984), (b) Tharmaraj *et al.* (1989), (c) Ghebrehiwet *et al.* (1988), (d) Saadullah (1985), (e) Perdok (1987), (f-h) Schiere *et al.* (1988), (i) Schiere *et al.* (1989). Except for Figure 2g where all data refer to TS rations, all symbols are as indicated in Figure a and i.

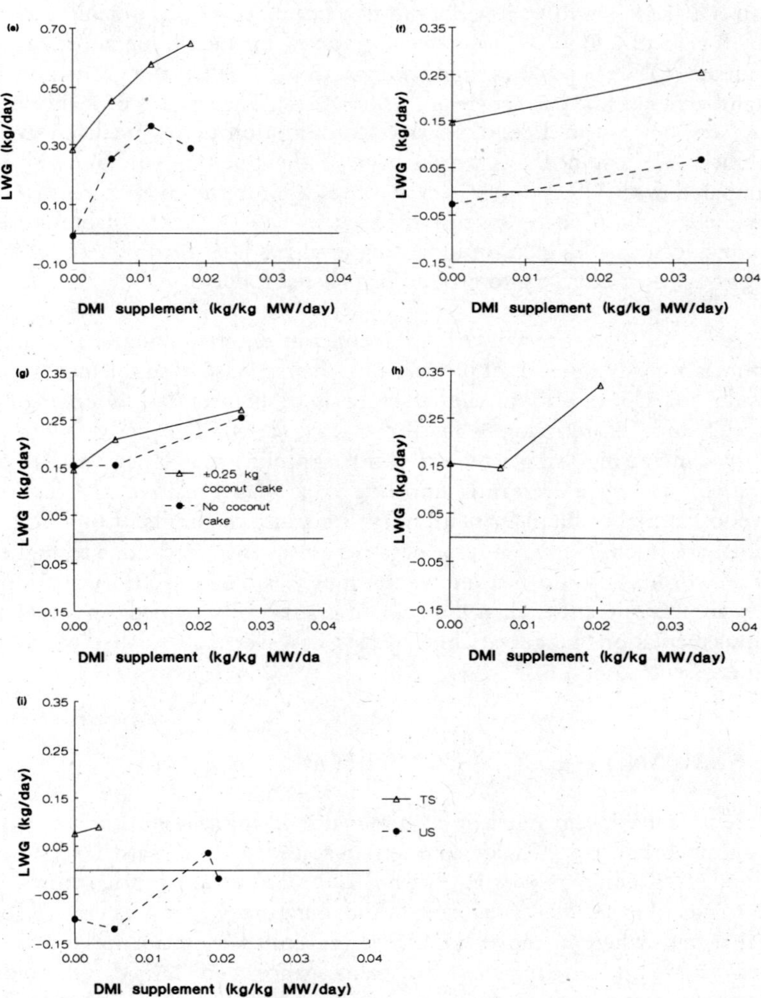

Figure 2.   Continued

Positive associative effects are often reported after the use of fishmeal supplement to animals fed with otherwise sufficient energy or with body reserves (Ørskov, 1981; Preston, 1985). Fishmeal contains a mix of minerals, growth factors and proteins of easier and slower degradability. Its quality is affected by origin and processing (Hussein and Jordan, 1991). Therefore, animal responses to fishmeal supplementation are unpredictable and cannot easily be ascribed to one of its components. Responses to slowly degradable protein such as fishmeal in the experiments of Saadullah (1985) can be caused by a high amino acid requirement of the young growing animals and/or by an effect of fishmeal supplementation on the extent and efficiency of rumen fermentation.

Associative effects can also be expected at high levels of nutrition. Fahmy and Sundstøl (1985) reported a decrease of *in sacco* digestibility at 60% concentrate levels in the ration for TS from 69.6% without supplement to 58.5% with supplement and for US from 46.5% to 42.5%. Badurdeen *et al.* (1994) report a low but not significant decrease of organic matter digestibility (OMD) for TS of 0.51 ($P > 0.10$) percentage point for each kilogram increase of organic matter intake (OMI) per 100 kg body weight (BW). For US the decrease was more drastic and significant i.e. 4.07 percentage points ($P < 0.01$). The effect is probably not due to associative effects, but to the effect of selective consumption of low quality leaves. If constant digestibilities are assumed - as we do below - the response to TS could be underestimated compared with US, another 'safety valve' against the overestimation of treatment effects in field application by extension services. After a review of literature, Aerts *et al.* (1984) concluded that an increase in feeding level has little or no effect on the digestibility of long roughages, contrary to ground or pelleted fodders.

Associative effects are less likely to occur if the ration contains different components, and if it is meant for animals at medium levels of production. The response to supplementation is consistently parallel for US and TS in the larger trials of Figures 2a-n. Converging response curves for US and TS are reported by Creek *et al.* (1984). This convergence is probably caused by increasing similarity of both rations at higher levels of supplementation, rather than by a decreasing nutritive value of TS relative to US. The response curves level off at higher supplementation levels because the limits of productive capacity of the animal are reached or because of negative associative effects due to higher supplementation levels. From this information, we cautiously assume for our calculations that linear responses are the rule rather than the exception, especially at medium levels of production and supplementation. Large variability exists however, and further work to improve prediction is essential.

REAL AND CALCULATED PERFORMANCE

When nutrient content of feeds and nutrient requirements of animals are known, it is possible to predict animal production (P_{calc}) within certain confidence intervals. To test the validity of prediction, one should compare P_{calc} with P_{real} for a number of trials, which is done here with the results of experiments reported in the literature. P_{calc} was calculated for dose response experiments, where a known DMI was multiplied by estimated nutrient contents and combined with requirements for maintenance and LWG. Of some experiments only the linear regression of growth on DMI was known (Ghebrehiwet *et al.*, 1988; Tharmaraj *et al.*, 1989), thereby artificially reducing variance in our calculations.

Considering the quality and detail of the available data, the only possible approach is to base calculations on TDN and total CP requirements.

The P_{real} relates quite linearly with the P_{calc} on a TDN basis (Figure 3). The best fit was obtained with high TDN requirements for maintenance and production and low TDN values of the straw. Low energy efficiency of straws was also observed by Perdok *et al.* (1982) and Khan and Davis (1981). The disappointing response to energy from straw based

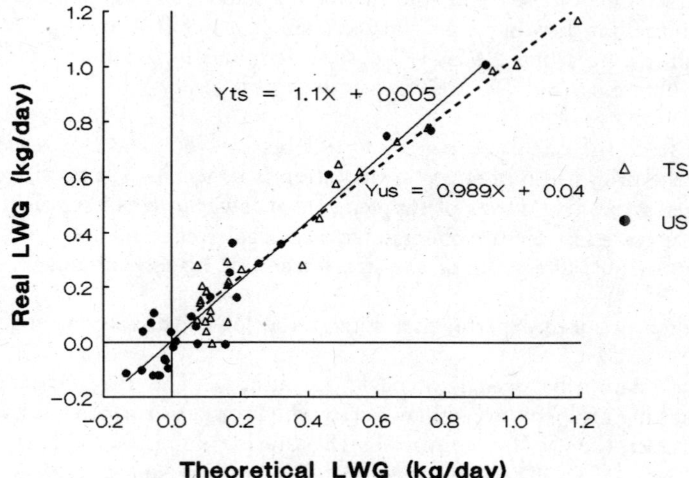

Figure 3. Review of dose-response curves and calculation of production based on TDN requirements. Sources: Creek *et al.* (1984), Saadullah (1985), Perdok (1987), Ghebrehiwet *et al.* (1988), Schiere and Wieringa (1989), Schiere *et al.* (1989), Tharmaraj *et al.* (1989).

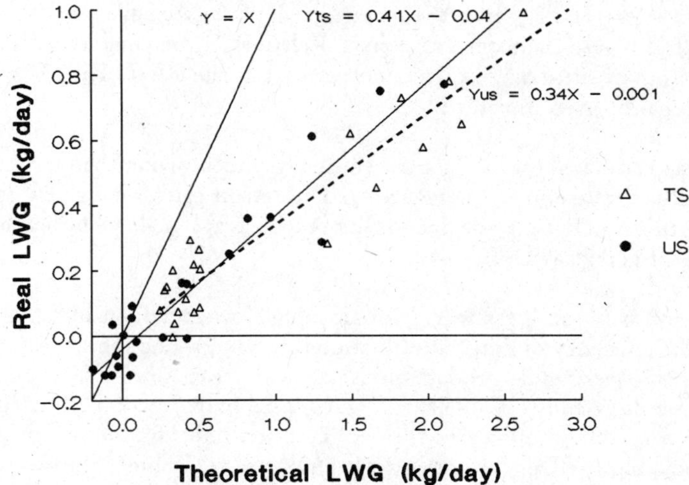

Figure 4. Review of dose-response curves and calculation of production based on CP requirements. Sources: Creek *et al.* (1984), Saadullah (1985), Perdok (1987), Ghebrehiwet *et al.* (1988), Schiere *et al.* (1989), Tharmaraj *et al.* (1989).

rations can be caused by the relative maturity of most experimental animals and/or environmental stress. There might also be an overestimation of nutritive value of the poorer rations in the TDN system, which is caused by a lower metabolisability of feeds containing high percentages of neutral detergent fibre (NDF) (Van Es, 1986) as a result of factors such as higher methane losses, more fermentation heat and possibly a poorer utilisation of fermentation end products such as volatile fatty acids.

The difference in slope of the regression line for P_{calc} and P_{real} of US and TS is small, and can result from the fact that they are estimated over slightly different ranges of LWG. The difference is almost nil when LWGs lower than zero and higher than 0.65 kg day^{-1} are excluded to compare TS and US at the same range of P_{real}.

The deviation from the calculated regression line between P_{calc} and P_{real} is highest at low LWGs where the proportion of straw in the ration is larger. This is possibly caused by:
- relative larger associative effects of supplements at lower levels of supplementation;
- a relatively larger effect of variable maintenance requirements;
- the variation of nutritive value of the straw exerts its biggest influence at high straw levels;
- the availability of more experimental data at the lower than at the higher range of production.
In all cases the relation between P_{calc} and P_{real} is positive. This is no great news, but it implies that ranking problems are unlikely even when the performance tends to be overestimated. In all cases, the LWG of animals on TS alone is around 100-150 g day^{-1} per animal for animals of 100-200 kg BW. This means that TS alone can support above maintenance levels of production in large ruminants. Small ruminants find it more difficult to maintain themselves at TS alone (De Jong and Van Bruchem, 1993, p.29).

Some of the trials used in Figure 3 (Saadullah, 1985; Perdok, 1987) have a much steeper relation between P_{real} and P_{calc} than the average of all trials. We have already speculated that associative effects cause such steep responses. Relatively lower maintenance requirements in theory than in practice may explain another part of the effect. They would imply that P_{calc} at maintenance overestimates P_{real}.

The P_{calc} values calculated on a CP basis were always much higher than the P_{real} though the pooled values still relate linearly (Figure 4). The reason cannot be given on the basis of available information, but may be due to factors such as the quality of the available protein or a shortage of energy (TDN).

The comparison between P_{calc} and P_{real} for lactating cows includes additional uncertainties concerning the efficiency of nutrient utilisation for milk production, butterfat content and pregnancy. Moreover, milk production is affected more than LWG by the previous production and disease history. Nell *et al.* (1986) used LCRF with TDN values of 0.37 for US and 0.45 for TS, and calculated that on TS only, milk production of around 2-4 kg should be possible for dairy animals with 350 kg BW. Table 5 shows milk yields of lactating cows and buffaloes corrected for liveweight changes and supplementation, thereby estimating an adjusted milk yield on TS only. It indicates that actual milk production varies considerably, although 2-4 kg milk seems to be a reasonable estimate for milk production

Table 5 Milk yield on TS only, extrapolated from low to medium supplementation in some experiments.

Milk yield (in kg with 4% fat)	Source
5.20	Van der Hoek *et al.*, 1989
2.87	Van der Hoek *et al.*, 1989
0.69	Van der Hoek *et al.*, 1989
2.42	Perdok *et al.*, 1982
3.53	Perdok *et al.*, 1982
0.26	Khan and Davis, 1981[a]
2.16	Khan and Davis, 1981[a]

[a] Fat content unknown, 4% fat is assumed.
Note: differences within experiments are related to the level of concentrate in the feed.

with these small animals on TS alone (Table 5). Such milk yields appear low, but they are quite common in tropical areas (Cunningham and Syrstad, 1987). Agrawal *et al.* (1989) reports that milk production of 6-7 kg is possible if TS is supplemented with 1 kg DM of legumes, but these authors do not give values on changes in liveweight.

CONCLUSIONS

Rations and technologies need to be pre-screened before they are tested or recommended in the field. Proper LCRF saves resources which otherwise would be necessary for animal response trials. The main technical problems for LCRF are the reliability of the data sets on nutrient requirements and values for nutritive value of different feeds.

Literature information on nutritive value and requirements for rice straw based rations shows a large variation, thus complicating LCRF. They include the effect of selective consumption, the differences of requirements and nutritive values between animals and straws, the associative effects and variable substitution rates.

Our comparisons between P_{calc} and P_{real} are based on high nutrient requirements and low TDN and CP values of the feed. This provides a safety margin where precise information is lacking. The relation between P_{calc} and P_{real} tends to be linear both on TDN and CP basis, but P_{calc} consistently overestimates P_{real}. This discrepancy can result from an overestimate of nutritive value of straws by using TDN and CP, as well as from an underestimate of the animal requirements. More complications arise from the effects of environmental stress, low energy availability from rumen degraded straw and/or a difference in genetic background of the animals in relation to the type of animals on which NRC standards are based. The data do not permit elaboration of the causes, but further work on those issues is highly relevant.

The difference between P_{real} and P_{calc} tends to increase at lower rates of growth and milk production. At such production levels the proportion of straw in the ration is relatively large, the variation between straws exerts a stronger effect, the effect of variable maintenance requirements is relatively large and associative effects are more likely.

Converging and levelling dose-response curves for TS and US at higher supplementation levels are likely. They are at least partly caused by increasing similarity between TS and US rations at high supplementation, and also because plateaus in animal production are reached.

The correlation between P_{real} and P_{calc} shows that LCRF is useful for ranking and *ex-ante* prediction of animal responses, but proper care needs to be taken not to overestimate P_{real}. The accuracy of LCRF should be increased by establishing more accurate data sets on the nutritive value and animal requirements. Special emphasis needs to be given to aspects of non-linear responses and the discrepancy between P_{real} and P_{calc}.

Feeding trials are indispensable but they should emphasise the measurement of OMI, OMD and rumen parameters in dose-response trials to allow better interpretation of results. Many supplementation levels - with fewer animals per treatment groups - seem preferable over a few levels of supplementation and many animals per treatment groups to quantify non-linear associative and substitutional effects. Such trials should include a zero-supplementation level for a better prediction and understanding of non-linear and associative effects.

ACKNOWLEDGEMENTS

Thanks are due to J. van Bruchem, B. Tolkamp, S.J. Oosting and S. Tamminga for their constructive comments, and to P. Martin for correcting the English.

REFERENCES

Aerts, J.V., De Boever, J.L., Cottyn, B.G., De Brabander, D.L. and Buysse, F.X., 1984. Comparative digestibility of feedstuffs by sheep and cows. Anim. Feed Sci. Technol., 12: 47.

Agrawal, I.S., Verma, M.L., Singh, A.K. and Pandey, Y.C., 1989. Feeding value of (urea) treated straws in terms of milk production. Indian J. Anim. Nutr., 6 (2): 89-96.

Agricultural Research Council, 1980. The Nutrient Requirements of Ruminant Livestock. Commonwealth Agricultural Bureaux (CAB), Farnham Royal, UK, 351 pp.

Badurdeen, A.L., Ibrahim M.N.M. and Schiere, J.B., 1994. Methods to improve utilization of rice straw: 2. Effect of different levels of feeding on intake and digestibility of untreated and urea ammonia treated rice straw. Asian-Australasian J. Anim. Prod. (accepted).

Barton, D., 1987. Draught animal power in Bangladesh. Ph.D. Thesis, School of Development Studies of the University of East Anglia, 250 pp.

Bhargava, P.K., Ørskov, E.R. and Walli, T.K., 1988. Rumen degradation of straw. 4. Selection and degradation of morphological components of barley straw by sheep. Anim. Prod., 47: 105-110.

Capper, B.S., 1988. Genetic variation in the feeding value of cereal straw. Anim. Feed Sci. Technol., 21: 127-140.

Chenost, M. and Reiniger, P. (eds.), 1989. Evaluation of Straws in Ruminant Feeding. Elsevier Applied Science, London, 182 pp.

Chuzaemi, S., Soebarinoto, S., Van Bruchem, J. and Prawirokusumo, S., 1992. *In sacco* digestibility as a predictor of digestibility and voluntary intake of rice straw. In: M.N.M. Ibrahim, R. De Jong, J. Van Bruchem and H. Purnomo (eds.), Proc. of the International Seminar on Livestock and Feed Development in the Tropics, Brawijaya University, Malang, Indonesia, October 1991, pp. 218-224.

Chuzaemi, S., Sulastri, R.D., Williams, A., Lammers-Weinhoven, T.S.C.W., Chen, X.B., Blümmel, Zemmelink, G., Prawirokusumo, S. and Van Bruchem, J., 1994. Effect of level of excess rice straw on intake and digestibility and rumen microbial protein synthesis in relation to retention time in the rumen of sheep. Anim. Feed Sci. Technol. (in preparation).

Coombe, J.B. and Tribe, D.E., 1962. The effects of urea supplements on the utilization of straw plus molasses diets by sheep. Aust. J. Agric. Res., 14: 70-91.

Cottyn, B.G., De Boever, J.L. and Vanacker, J.M., 1989. *In vivo* digestibility measurements of straws. In: M. Chenost and P. Reiniger (eds.), Evaluation of Straws in Ruminant Feeding. Elsevier Applied Science, London, pp. 36-46.

Creek, M.J., Barker, T.J. and Hargus, W.A., 1984. The development of a new technology in ancient land. World Anim. Rev., 51: 12-20.

Cronjé, P.B., 1990. Supplementary feeding in ruminants - a physiological approach. S. Afr. J. of Anim. Sci., 20(3): 110-117.

Cunningham, E.P. and Syrstad, O., 1987. Crossbreeding *Bos indicus* and *Bos taurus* for milk production in the tropics. FAO Anim. Prod. Health Pap. 68, FAO, Rome. 90 pp.

De Jong, R. and Van Bruchem, J., 1993. Utilization of crop residues and supplementary feeds in tropical developing countries. Final Report for Commission of the European Communities. Programme: Science and Technology for Development, no. TS2-0091-NL, 96 pp.

Djajanegara, A. and Doyle, P.T., 1989. Urea supplementation compared with pretreatment. 1. Effects on intake, digestion and liveweight change by sheep fed on rice straw. Anim. Feed Sci. Technol., 27: 17-30.

Doyle, P.T. and Panday, S.B., 1990. The feeding value of cereal straws for sheep. III. Supplementation with minerals or minerals and urea. Anim. Feed Sci. Technol., 29: 29-43.

Doyle, P.T. and Pearce, G.R., 1985. Processing methods to improve nutritive value of rice straw. In: Efficient Animal Production for Asian Welfare, Proc. of the 3rd A.A.A.P. Animal Science Congress, 6-10 May 1985, Vol. 1, Seoul, Korea, pp. 82-98.

Doyle, P.T., Devendra, C. and Pearce, G.R., 1986. Rice straw as a feed for ruminants. IDP, Canberra, A.C.T., 117 pp.

Doyle, P.T., Pearce, G.R. and Djajanegara, A., 1987. Intake and digestion of cereal straws. In: T.F. Reardon, J.L. Adam, A.G. Campbell and R.M.W. Sumner (eds.), Proc. of the 4th A.A.A.P. Animal Science Congress, Hamilton, New Zealand, pp. 59-62.

Durand, M., 1989. Conditions for optimizing cellulolytic activity in the rumen. In: M. Chenost and P. Reiniger (eds.), Evaluation of Straws in Ruminant Feeding. Elsevier Applied Science, London, pp. 3-18.

Elliot, R.C., Mills, W.R. and Reed, W.D., 1966. Survival feeding of afrikaner cows. Rhod. Zambia Malawi J. Agric. Res., 4: 69-75.

Fahmy, S.T.M. and Sundstøl, F., 1985. The degradability of untreated and chemically-treated barley straw and of grass silage as influenced by the ration composition. Z. Tierphysiol., Tierernähr. Futtermittelkd., 53: 34-42.

Faverdin, P., Dulphy, J.P., Coulon, J.B., Vérité, R., Garel, J.P., Rouel, J. and Marquis, B., 1991. Substitution of roughage by concentrates for dairy cows. Livest. Prod. Sci., 27: 137-156.

Forbes, J.M., 1988. The prediction of voluntary intake by the dairy cow. In: P.C. Garnsworthy (ed.), Nutrition and Lactation in the Dairy Cow. Butterworths, London, pp. 294-312.

Fox, D.G., Sniffen, C.J. and O'Connor, J.D., 1988. Adjusting nutrient requirements of beef cattle for animal and environmental variations. J. Anim. Sci., 66: 1475-1495.

Frisch, J.E. and Vercoe, J.E., 1978. Utilizing breed differences in growth of cattle in the tropics. World Anim. Rev., 25: 8-12.

Ghebrehiwet, T., Ibrahim, M.N.M., Schiere, J.B., 1988. Response of growing bulls to diets containing untreated or urea treated rice straw with rice bran supplementation. Biol. Wastes, 25(4): 269-280.

Hussein, H.S. and Jordan, R.M., 1991. Fish meal as a protein supplement in ruminant diets: a review. J. Anim. Sci., 69: 2147-2156.

Hvelplund, T., 1989. Protein evaluation of treated straws. In: M. Chenost and P. Reiniger (eds.), Evaluation of Straws in Ruminant Feeding. Elsevier Applied Science, London, pp. 60-74.

Ibrahim, M.N.M., Tamminga, S. and Zemmelink, G., 1989. Effect of urea treatment on rumen degradation characteristics of rice straws. Anim. Feed Sci. Technol., 24: 83-95.

Jayasuriya, M.C.N. and Perera, H.G.D., 1982. Urea ammonia treatment of rice straw to improve its nutritive value for ruminants. Agric. Wastes, 4: 143-150.

Kellaway, R.C. and Leibholz, J., 1983. Effect of nitrogen supplements on intake and utilization of low quality forages. World Anim. Rev., 48: 33-37.

Khan, A.K.M.N. and Davis, C.H., 1981. Effect of treating paddy straw with ammonia (generated from urea) on the performance of local and crossbred lactating cows. In: M.G. Jackson, F. Dolberg, C.H. Davis, Mozammel Haque and M. Saadullah (eds.), Maximum Livestock Production from Minimum Land. Proc. of a Seminar, 2-5 February 1981, Bangladesh, pp. 168-180.

Laurenz, J.C., Byers, F.M., Schelling, G.T. and Greene, L.W., 1991. Effect of season on the maintenance requirements of mature beef cows. J. Anim. Sci., 69: 2168-2176.

Leng, R.A., 1990. Factors affecting the utilization of 'poor quality' forages by ruminants particularly under tropical conditions. Nutr. Res. Rev., (3): 277-303.

Mould, F.L., Ørskov, E.R. and Mann, S.O., 1983. Associative effects of mixed feeds. I. Effects of type and level of supplementation and the influence of the rumen fluid pH on cellulolysis *in vivo* and dry matter digestion of various roughages. Anim. Feed Sci. Technol., 10: 15-30.

National Research Council, 1976. Nutrient requirements of Domestic Animals No.4, Beef cattle, 5th revised edn. National Academy of Sciences, Washington, DC, 56 pp.

National Research Council, 1978. Nutrient requirements of Domestic Animals No.3, Dairy cattle 5th revised edn. National Academy of Sciences, Washington, DC, 76 pp.

National Research Council, 1984. Nutrient requirements of Beef cattle, 6th revised edn. National Academy of Sciences, Washington, DC, 90 pp.

National Research Council, 1987. Predicting feed intake of food processing animals. National Academy Press, Washington, DC, 85 pp.

National Research Council, 1988. Nutrient requirements of Dairy cattle, 6th revised edn. National Academy of Sciences, Washington, DC, 157 pp.

Navaratne, H.V.R.G., Ibrahim, M.N.M. and Schiere, J.B., 1990. Comparison of four techniques for predicting digestibility of animal feeds. Anim. Feed Sci. Technol., 29: 209-221.

Nell, A.J., Schiere, J.B. and Ibrahim, M.N.M., 1986. Economic evaluation of urea ammonia treated straw as cattle feed. In: M.N.M. Ibrahim and J.B. Schiere (eds.), Rice Straw and Related Feeds in Ruminant Rations. Proc. of an Int. Workshop, Kandy, Sri Lanka, 24-28 March 1986, Straw Utilization Project (SUP), Kandy, Sri Lanka, pp. 164-170.

Ogink, N.W.M., 1993. Genetic size and growth in goats. Ph.D. Thesis, Wageningen Agricultural University, Wageningen, 171 pp.

Oosting, S.J., 1993. Wheat straw as ruminant feed. Effect of supplementation and ammonia treatment on voluntary intake and nutrient availability. Ph.D. Thesis, Wageningen Agricultural University, Wageningen, 232 pp.

Ørskov, E.R., 1981. Feeding young ruminants in the tropics. In: M.G. Jackson, F. Dolberg, C.H. Davis, M. Haque and M. Saadullah (eds.), Maximum Livestock Production from Minimum Land. Proc. of a Seminar, 2-5 February 1981, Bangladesh, pp. 264-270.

Perdok, H.B., 1987. Ammoniated rice straw as a feed for growing cattle. Ph.D. Thesis, University of New England, Armidale, N.S.W.

Perdok, H.B., Thamotharam, M., Blom, J.J., Van Den Born, H. and Van Veluw, C., 1982. Practical experiences with urea-ensiled straw in Sri Lanka. In: T.R. Preston, C.H. Davis, F. Dolberg, M. Haque and M. Saadullah (eds.), Maximum Livestock Production from Minimum Land, Proc. of the 3rd Seminar, Bangladesh Agricultural Research Institute, Joydebpur, Bangladesh.

Powell, M.J., 1985. Yields of sorghum and millet and stover consumption by livestock in the sub humid zone of Nigeria. Trinidad, Trop. Agric., 62: 77-81.

Preston, T.R., 1985. Validity of feeding standards and development of feeding systems based on crop residues and agro-industrial by-products. In: T.R. Preston, Vappu L. Kossila, J. Goodwin and S.B. Reed (eds.), Better Utilization of Crop Residues and By-products in Animal Feeding: Research Guidelines. 1. State of Knowledge. FAO Anim. Prod. Health Pap. 50, FAO, Rome, pp. 197-213.

Rai, S.N. and Gupta, B.N., 1989. Effect of long term feeding of urea (ammonia) treated wheat straw on intake, nutrient utilization, N metabolism and economics of feed cost in yearling murrah buffalo bulls. Buff. J. Thailand (accepted).

Reed, J.D., Capper, B.S. and Neate, P.J.H. (eds.), 1988. Plant breeding and the nutritive value of crop residues. Proc. of a Workshop, ILCA, Addis Ababa, Ethiopia, 7-10 December 1987, ILCA, Addis Ababa, Ethiopia, 334 pp.

Rohr, K., Lebzien, P., Schafft, H. and Schulz, E. 1979. Prediction of duodenal flow of non-ammonia nitrogen and ammonia acid nitrogen in dairy cows. Livest. Prod. Sci., 14: 29-40.

Roth, L.B., Rouquette, F.M. and Ellis, W.C., 1990. Effect of herbage allowance on herbage and dietary attributes of coastal bermuda grass. J. Anim. Sci., 68: 193-205.

Saadullah, M., 1985. Supplementing urea-treated rice straw for native cattle in Bangladesh. In: M. Wanapat and C. Devendra (eds.), Relevance of Crop Residues in Developing Countries. Proc. of an Int. Workshop, Khon Kaen, Thailand, 315-329.

Schiere, J.B. and De Wit, J., 1993. Feeding standards and feeding systems. Anim. Feed Sci. Technol., 43: 121-134.

Schiere, J.B. and Ibrahim, M.N.M., 1989. Feeding of urea-ammonia treated rice straw. A compilation of miscellaneous reports produced by the Straw Utilization Project (Sri Lanka). PUDOC, Wageningen, 125 pp.

Schiere, J.B. and Nell, A.J., 1993. Feeding of urea treated straw in the tropics. I: A review of its technical principles and economics. Anim. Feed Sci. Technol., 43: 135-147.

Schiere, J.B. and Wieringa, J., 1989. Overcoming the nutritional limitations of rice straw for ruminants. 2. Response of growing Sahiwal and local cross heifers to urea upgraded and urea supplemented straw. Asian-Australasian J. Anim. Sci., 1: 213-218.

Schiere, J.B., Kumarasuntharam, V.R., Sewalt, V.J.H. and Brouwer, B., 1988. Overcoming the nutritional limitations of rice straw for ruminants. 3. Urea ammonia upgrading of straw and supplementation with rice bran and coconut cake for growing bulls. Asian-Australasian J. Anim. Sci., 1: 213-218.

Schiere, J.B., Ibrahim, M.N.M., Sewalt, V.J.H. and Zemmelink, G., 1989. Response of growing cattle given rice straw to lickblocks containing urea and molasses. Anim. Feed Sci. Technol., 26: 179-189.

Schneider, B.H. and Flatt, W.P., 1972. The Evaluation of Feeds through Digestibility Experiments. University of Georgia Press, Athens, GA, 423 pp.

Seanger, P.F., Lemenager, R.P. and Hendrix, K.S., 1983. Effects of anhydrous ammonia treatment of wheat straw upon *in vitro* digestion, performance and intake by beef cattle. J. Anim. Sci., 56(1): 15-20.

Soebarinoto, S., Chuzaemi, S. and Van Bruchem, J., 1990. Nutritive value of rice straw varieties in Indonesia as related to location and urea treatment. In: H. Kuil, R.W. Paling and J.E. Huhn (eds.), Livestock Production and Diseases in the Tropics. Proc. 6th Int. Conference of Institutes for Tropical Veterinary Medicine. University of Utrecht, Utrecht.

Sundstøl, F. and Coxworth, E.M., 1984. Ammonia treatment, Chapter 7. In: F. Sundstøl and E. Owen (eds.), Straw and Other Fibrous By-products as Feed. Elsevier, Amsterdam, pp. 196-247.

Tamminga, S. and Van Hellemond, K.K., 1977. The protein requirements of dairy cattle and developments in the use of protein, essential amino acids and non-protein nitrogen in the feeding of dairy cattle. In: Protein and Non Protein Nitrogen for Ruminants. Pergamon Press, Oxford. pp. 9-32

Taylor, St. C.S., 1985. Use of genetic size-scaling in evaluation of animal growth. J. Anim. Sci., 61 (Suppl.) 2.

Tharmaraj, J., Van Der Hoek, R., Sewalt, V.J.H. and Schiere, J.B., 1989. Overcoming the nutritional limitations of rice straw for ruminants. 4. Urea ammonia treatment and supplementation with *Gliricidia maculata* for growing Sahiwal bulls. Asian-Australasian J. Anim. Sci., 2(2): 85-90.

Tolkamp, B.J. and Ketelaars, J.J.M.H., 1992. Toward a new theory of feed intake regulation in ruminants 2. Cost and benefits of feed consumption: an optimization approach. Livest. Prod. Sci., 30(4): 297-317.

Ulhas, H. Prabhu, Subba Rao, A., Sampath, S.R. and Schiere, J.B., 1988. Effect of allowance on the intake and digestibility of forages. In: Kiran Singh and J.B. Schiere (eds.), Fibrous Crop Residues as Animal Feed. Aspects of Treatment, Feeding, Nutrient Evaluation, Research and Extension. Proc. of an Int. Workshop, 27-28 october 1988, National Dairy Research Institute, Bangalore, Indo-Dutch Project on Bioconversion of Crop Residues.

Van Bruchem, J., Oosting, S.J., Bongers, D.L.J.G.M., Lammers-Wienhoven, T.S.C.W. and Gurses, C., 1992. Treatment of wheat straw with ammonia: Effect of energy and protein supplements on intestinal available amino acids in sheep. In: M.N.M. Ibrahim, R. De Jong, J. Van Bruchem and H. Purnomo (eds.), Livestock and Feed Development in the Tropics. Proc. of the Int. Seminar, Brawijaya University, Malang, Indonesia.

Van Der Hoek, R., Muttetuwegama, G.S. and Schiere, J.B., 1989. Overcoming the nutritional limitations of rice straw for ruminants. 1. Urea ammonia treatment and supplementation of straw for lactating Surti buffaloes. Asian-Australasian J. Anim. Sci., 1: 201-208.

Van Es, A.J.H., 1986. Energy utilization of low digestible carbohydrates. Br. J. Nutr., 55: 545-554.

Van Soest, P.J., 1982. Nutritional Ecology of the Ruminant. Cornell University Press, Ithaca, NY, 243 pp.

Wahed, R.A., Owen, E., Neate, M. and Hosking, B.J., 1990. Feeding straw to small ruminants: the effect of amount offered on intake and selection of barley straw by goats and sheep. Anim. Prod., 52: 283-289.

Walli, T.K., Ørskov, E.R. and Bhargava, P.K., 1988. Rumen degradation of straw. 3. Botanical fractions of two rice straw varieties and effect of ammonia treatment. Anim. Prod., 46: 347-352.

Zemmelink, G., 1980. Effect of selective consumption on voluntary intake and digestibility of tropical forages. Doctoral Thesis, Agric. Res. Rep. 896, Centre for Agricultural Publishing and Documentation (PUDOC), Wageningen, 100 pp.

Zorilla-Rios, J., Horn, G.W. and McNew, R.W., 1989. Effect of ammoniation and energy supplementation on the utilization of wheat straw by sheep. Anim. Feed Sci. Technol., 22: 305-320.

Section 5

THE SHAPE OF SYSTEMS

ah love, could thou and I with fate conspire
to grasp this sorry state of things entire,
would we not shatter it to bits and then
remould it nearer to our heart's desire
> Omar Khayyám, 11th century AD, in the "Rubaiyat"; Harrap edition, Harrap Ltd, London.

'where fodder is scarce, cows should only be allowed to calve every second year, particularly
when cows are used for farm work, to enable the cow to have an ample supply of nourishment
for her calf and to save her the double burden of work and pregnancy'
> Columella, a Roman writer on farming and farming systems in the first century A.D. (quoted by
> K.D. White, 1970, Roman Farming, Thames and Hudson, London.)

Chapter 5.1

FEEDING STANDARDS AND FEEDING SYSTEMS[1]

J.B. Schiere and J. De Wit

SUMMARY

Feeding standards are commonly understood to be developed in and for feeding systems of temperate countries. The application of these feeding standards is often challenged, particularly for the tropics, but the discussion about their relevance needs clear definitions and statements of objective to be fruitful. First of all, this paper defines and discusses the concepts of feeding standards in relation to the objectives of their use. Secondly, it suggests that for development purposes the classification of temperate and tropical be replaced by high and low input systems. In the high input systems, it is mainly the feeds that are adapted to the animal; in the low input systems, it is often the animal that needs to be adapted to the feed. It is then argued that the relevance of feeding standards depends on: (a) the technical validity of the datasets, and (b) the socio-economic aspects of their application. So-called low animal production levels in 'low input' systems are not only caused by feed shortages in terms of quality and quantity, but also by socio-economic considerations, rather than by an ignorance of feeding standards on the part of the farmer. Scientists and policy makers often misinterpret farmers' production goals, leading to improper application of standards. The technical reliability of standards can still be increased but the desired precision depends on the purpose for which they are used. Prediction of accurate absolute values in variable conditions will be difficult for many years to come, but this need not delay the use of recent knowledge on standards in order to better understand systems and farmers' reasoning as well as to set research and extension priorities.

---

[1] Published in Animal Feed Science and Technology, 43(1993): 121-134.

INTRODUCTION

The use of feeding standards can be challenged in general, but more so when they are developed for one type of feeding system and applied in another. The use of temperate standards for tropical systems is a particular point of contention (McGraham, 1983; Leng, 1990). Preston (1985) doubts the reliability of traditional concepts for systems in the tropics, but not the relevance of feeding standards *per se*, while Jackson's (1981a) question *'who needs feeding standards'*, is mainly based on doubts about socio-economic applicability. In spite of these arguments, feeding standards can be useful for *ex-ante* testing of innovations in farming systems where it is impractical to test each feed combination with animal trials (Potts, 1982; Chesworth *et al.*, 1989; Schiere and Nell, 1993). Modelling animal responses and feed requirements can also identify the gaps in our knowledge about the data sets (Gill, 1991) as well as enabling us to understand farmers' reasoning (De Wit *et al.*, 1993). Both critics and advocates of feeding standards can be right, as long as the purpose and conditions of their use are not well defined.

This article first defines feeding standards and their objectives. Secondly, it reconsiders the distinction between tropical and temperate farming systems. Thirdly, the applicability of feeding standards is discussed with regard to the technical validity of the data sets and socio-economic aspects of their application. The paper relates mainly to systems with ruminants on roughage-based diets, but the point is also valid for other animals and feeding systems.

FEEDING STANDARDS: CONCEPT AND USE

Feeding standards are discussed in many scientific meetings and tropical/temperate classrooms. Their relevance for farmers is often disputed, but it is then forgotten that even farmers apply feeding standards. The application of feeding standards by farmers is of varying intensity, not necessarily in the form of printed tables and often with measures and objectives that differ from those of researchers. However, even under low input scavenging or pastoral conditions, the farmers use their informal feeding standards to guide animals towards better grazing grounds or away from unfavourable areas.

Many scientists and development workers are reluctant to apply the 'science' of standards to farmers' conditions. The reason for this include unfamiliarity with the concepts; other arguments are as follows:

(1) The fear of incorrect and static use of book values in situations for which they were not developed. Shah *et al.* (1980) formulated a common but incorrect sentiment on the use of feeding standards: *'scientific feeding rates maximize milk production, they do not maximize the farmers returns from milk production'*. The statement implies that feeding standards are a static rule to be applied for high production levels, rather than a flexible tool to assist common sense.

(2) The use of standards is often assumed to refer to animals and feeds alone, with insufficient attention to other components of the system, i.e. crops, eroding grazing areas, etc. Instead of applying the standards to more general problems, they are abandoned altogether.

(3) The application of feeding standards from one system to another can lead to the improper classification of feeds as good or bad, without realizing that what is good in one

system may be bad in another. For example, straw is a bad feed for dairy animals in many high input systems, but it is quite a reasonable feed to maintain animals through a dry season. Straw is actually a valuable source of fibre for urban dairies where large quantities of concentrate are fed. In some urban feed/fodder markets of the tropics the price per unit of dry matter for straw exceeds that of concentrate!

(4) The imprecision of performance prediction from standards in variable situations makes many workers forget the usefulness of standards to predict general trends in feeding systems (McGraham, 1983; Preston, 1985). This leads to endless testing of rations on farm or on station conditions. Such testing, however, cannot represent real situations, nor can it predict accurate responses owing to a lack of animals, etc.

Confusion about the need for feeding standards can be overcome by definition of the terminology and objectives of their use. The terms 'nutrient requirements' and 'feeding standards' can be used interchangeably. Our definition is based on that proposed by Crampton and Harris (1968):

'feeding standards are data sets which record what is believed to be the need of a specific animal for one or more of the recognized feed components for a given period of time'.

The original word 'daily' is replaced with 'for a given period of time', to include time effects of compensatory gain or negative carry-over from previous poor nutrition. Such data sets need not necessarily be recorded in terms of energy and protein values. Standards that exist in many farmers' minds include criteria of feed evaluation, that for example relate to the effect of such feed on butterfat content or dung consistency (Rangnekar, 1993), but these criteria represent aspects of nutritive value nevertheless. Further refinements to the definition of feeding standards are possible, e.g. the need of a specific animal can be replaced by the need for a group of animals etc., but semantics will not serve the point of this paper. Feed analysis can be carried out by farmers as well as in a laboratory; either way it serves to establish the standards which in turn serve to establish feeding (or rationing) systems, to be defined as 'the method of feeding, in terms of allocation of quantity and quality of feedstuff(s) over time and animals, guided by a combination of farmers objectives, the feed availability and the animal'.

The usefulness and desired accuracy of feeding standards depends on the context and purpose for which they are used. Farmers, scientists and policy makers that develop their standards in isolation are likely to misunderstand each other. It should also be clear that the use of farmers and scientifically developed feeding standards are complementary rather than mutually exclusive. Scientists and policy makers often fail to understand farmers' values and priorities, particularly in low input systems. Much development policy operates on the assumption that high individual productions per animal implies high income. This paradigm is not new, as is shown by the following quotation from Jackson (1981b):

'the Royal Commission on Agriculture of 1927 first set forth the objectives of livestock development in India, namely increased production per animal and a reduction in animal numbers'.

That paradigm requires that feeds are combined to suit the purpose of high production per animal, i.e. an approach that is only appropriate for high input systems, as explained below. This does not, however, imply that the feeding standards (data sets) from high input systems are useless in low input systems. The discussion on feeding standards should concern the validity of the data sets and their application, not the principle of their existence and use. To discard the concept of feeding standards for its shortcomings would be to 'throw the baby out with the bath water'. For a better discussion of the application of standards we need first to reassess the traditional classification of tropical and temperate farming systems. The technical validity of the feeding standards will then be reviewed, together with issues of their application in different socio-economic conditions.

THE CLASSIFICATION OF SYSTEMS

The term 'tropical' generally but imprecisely implies concepts such as underdevelopment, inefficiency, non-Western, subsistence and tradition, whereas 'temperate' implies the opposite. Such a distinction is inadequate for a discussion on the usefulness of feeding standards for the following reasons:

(1) In physical terms, the tropics include snowcapped mountains, deserts, irrigated fertile lands and mangrove forest. In fact, it can be said that the 'tropics' include 'temperate systems', but that the reverse is not true.

(2) In socio-economic terms, the tropics include the entire range of farming systems between highly commercialised pig, poultry and dairy production on the one hand and subsistence livestock production on the other. Temperate systems, when implicitly equated with high input systems, show smaller variation of systems, because the high usage of inputs such as fertiliser, irrigation, purchased feeds etc., tends to mask differences between or within systems.

Input is the key word that we believe to be (incorrectly) implied in the terms tropical/temperate, or developing and developed. The intricacies of the resource/input terminology deserve further definition, but this is outside the scope of this paper. Let it suffice here, for development purposes, to use the term high and low input systems rather than temperate and tropical. High input in terms of feed use implies access to an unlimited range of high quality feedstuffs and a well developed market for the produce. Low input implies that for animal feed, producers have to rely on what is locally and seasonally available, e.g. roadside, forest or range grazing and crop residues, and often accept relatively low prices for their produce.

One essential difference between high and low input systems is that the manager (farmer) of a low input system adjusts the type and level of production to the feed availability, whereas in the high input system the inputs are adjusted to the desired level of production. In other words, a high input farmer buys feeds to increase the individual production of the purebred animal, whereas the low input farmer uses crossbred animals instead of a purebred high yielder that would not survive on the existing feed supplies. Low input systems adjust to periodic feed scarcity by allowing animals to reduce their daily gain, or even to lose weight. The purchase of feed in such systems is not a financially attractive option. Both low and high input systems occur in temperate as well as tropical regions, and both require the use of feeding standards, whether applying the data sets of farmers or those of scientists. Feeding standards for low input systems have been developed for temperate systems by the National Research Council (NRC) and the Agricultural Research Council (ARC), and have been reviewed for survival feeding by Barker and Stoate (1969), Oddy (1978) and Cronjé (1990). Obviously, standards based on data sets and concepts for one system cannot blindly be applied in another. We agree that tropical feeds are different from temperate feeds, but we maintain that the classification tropics/temperate diverts attention from the real issues, i.e. those of input availability and use.

TECHNICAL RELIABILITY OF FEEDING STANDARDS

Improvement of feed evaluation and estimates of requirements have been attempted since the Hay Value was conceived by Thaer (De Boer and Bickel, 1988; Baldwin and Hanigan, 1990). Even major temperate feed evaluation systems differ in concept and values, such as those of NRC and ARC. A reasonable agreement between methods for high input conditions is apparent from De Boer and Bickel (1988), but remarkable differences in response to feeds of the same apparent nutritive value were demonstrated by Preston (1985) and Leng (1990), particularly for cases such as (supplementation of) fibrous crop residues in low input systems.

Accuracy of prediction by ration formulation is likely to be less in the low input systems because of the inherent higher variability of such systems, expressed in physiological status and history of the animals, highly variable feed on offer, selective consumption and composition of the produce. Such issues have been reviewed for straw-based rations by Schiere and De Wit (1995). They conclude that calculated predictions based on NRC standards overestimate the actual responses but that ranking of responses is possible. The variability of low input systems contrasts strongly with the precision in performance prediction of up to 1-3% obtained by Rayburn and Fox (1990) for standardised high input feedlot conditions in the USA. Those conditions are not only less variable, the Rayburn-Fox models even include stress factors such as thickness of the mud in the cattle yard!

Difficulty in predicting animal response under low input conditions may also be caused by the following factors:

(1) Environmental stress, diseases, starvation and differences between and within breeds that affect (maintenance) requirements have all been cited as causing problems in predicting animal response under low input conditions (Elliot *et al.*, 1966; Frisch and Vercoe, 1978; Rayburn and Fox, 1990; Birkelo *et al.*, 1991; DiConstanzo *et al.*, 1991; Laurenz *et al.*, 1991). Variable maintenance requirements affect the accuracy of predicted performance more at low than at high productions levels.

(2) The feeding value of graminaceous feeds in tropical climatic conditions is less than that of fodder grown under well managed temperate conditions (Dirven, 1977) and the quality of feeds in low input systems varies more often, owing to larger variations in water availability, fertilization, soil type, climate, etc.

(3) The secondary effects of feeding regimes on (re)production or compensatory growth (O'Donovan, 1984; Robinson, 1990) also contribute to difficulty in predicting animal response under low input conditions.

Further improvement of feeding standards is possible and necessary, but important developments have taken place during the last decades. These include the following:

(1) Systems of protein and energy evaluation have been refined and the importance of their interrelation is increasingly understood (Balch and Van Es, 1986).

(2) Proximate analysis is supplemented by the Van Soest fibre analysis and measurements of *in vitro* digestibility and rates of degradation (Chenost and Reiniger, 1989).

(3) Tables of tropical feed values have become available to supplement the values from temperate literature. Examples are the tables from McDowell *et al.* (1974), Gohl (1981), Kearl (1982), Ibrahim (1988), and Ranjhan (1991) though their application still requires good judgement because of large within feed class variability to location or seasons.

(4) Values and discussions of dry matter intake (DMI) were absent in many early handbooks on animal nutrition. This is more serious where lower feed quality depresses intake, aggravating the overestimation of production responses. McDonald *et al.* (1973) is one of the first standard texts to mention intake. Theoretical understanding of DMI prediction is growing (Forbes, 1986; NRC, 1987; Owens, 1987; Ørskov *et al.*, 1988; Ketelaars and Tolkamp, 1991). Tables with DMI values have recently become available (Doyle *et al.*, 1986; Prasad *et al.*, 1993; Schiere and De Wit, 1995). Recent feed balances and ration formulation for tropical and low input conditions incorporate DMI constraints in the calculations (Zemmelink *et al.*, 1992; Schiere and Nell, 1993). A point to consider here is the need to distinguish between biologically determined maximum values of DMI, and the actual DMI which is influenced by economics and farmers' practices.

(5) Relations between feed on offer and quality of feed consumed are recognized and better understood (Zemmelink, 1980; Wahed *et al.*, 1990).

(6) Most temperate standards implicitly assumed that nutrient concentrations in the feed are adequate to support good rumen function, but this assumption is not valid for situations of overmature grass, where animals 'starve in a sea of plenty' (Altona, 1966). The importance of associative effects is increasingly recognized though the effects are still difficult to predict (Cronjé, 1990; Leng, 1990; Schiere and De Wit, 1995).

(7) Computing equipment facilitates least cost ration formulation (LCRF), and can include DMI limitations or substitution rates etc. in sensitivity analyses (Nell *et al.*, 1986; Chesworth *et al.*, 1989). Ration formulation suffers more from a lack of reliable data than from the versatility of the software. Feed companies all over the world use these techniques for commercial purposes, proving that they must have some value, particularly for high input poultry, pigs and dairy systems.

SOCIO-ECONOMIC ASPECTS OF THE USE OF FEEDING STANDARDS

Even if feeding standards were technically perfect, their application needs to be based on local and seasonal feed availability, as well as on the value of produce. Farmers in both high and low input systems can be assumed to produce for profit, but economics are not the same everywhere. Differences in resource allocation, demand patterns, social relations, personal preferences/job satisfaction, even within families and between gender groups, result in a vast array of (multiple) production goals and systems (Harris, 1965; Crotty, 1980; Behnke, 1985; Mace and Houston, 1989; Shanti George, 1991). Produce prices and labour costs differ even between categories of farmers within the same geographical district (Patel *et al.*, 1977). Risk factors, or low fixed costs in combination with low produce prices force farmers to accept low levels of production per animal (Doyle, 1974).

In many farming systems, farm production levels are limited by production quota, whether explicitly in the high input systems of the EC or implicitly in low input systems under poor marketing conditions of the tropics. Income in the case of production quota cannot be increased by a higher volume of production, only by a reduction in the cost per unit produced (Oscarsson, 1975; Welsch, 1975; Kristensen and Thysen, 1991). Whether such a reduction is achieved by increased production per animal or by a reduction in the price of feed depends mainly on the fixed costs of animal maintenance which again are very system specific. Multipurpose production objectives (meat, milk, young stock, dung, wealth, status and security) and hidden costs (disease risks, environmental side-effects such as erosion of common grazing lands, stress of entrepreneurship) make it misleading to estimate profitability only in terms of a single output and input. These conditions allow no simple application of the earlier statement by the Royal Commission, even less so when animal production is secondary or integrated with cropping in mixed farming systems (Kidane, 1984). These micro-economic issues still ignore the need to reorient economic thinking for sustainability, for example to avoid externalization of hidden production costs (Conway and Barbier, 1990). Economics are difficult to estimate in systems where milk production is a by-product of the production of bullocks for draught, where children tend the animals or where feed is obtained from communal grazing.

Maximization of combined crop-livestock production can require in some farming systems that one or both components of the system have to sacrifice individual productions for the benefit of the total. Examples of this are where farmers keep lower producing animals to utilize so-called low quality feed that is available on farm, or accept lower grain yields when high grain yields do not correlate with a high income, in farming systems where the stover has a relatively high value for animal feed (Nordblom and Halimeh, 1982; Kelley *et al.*, 1991; De Wit *et al.*, 1993).

A commonly accepted principle requires the minimization of the feed conversion ratio (FCR), i.e. units of feed consumed per unit produced. This, however, is only relevant when all feed classes are equally scarce (expensive). The low FCR in Table 1 is clearly not financially attractive, particularly because the price of liveweight gain (LWG) in that system was less than 10 Rs kg^{-1}, i.e. less than the cost of liveweight gain under the conditions of lowest FCR. Moreover, the aggregate FCR conversion (Table 1) shows a different optimum production than the concentrate conversion ratio (CCR) alone. Where feed quality (e.g. concentrate availability) is limiting, it might therefore be advisable to aim for low values of CCR at the expense of higher aggregate FCR values. The use of feeding standards can assist in outlining responses and optimum levels of production as shown in the example of Table 2 where the least cost ration at a given production level is not necessarily the financially most attractive ration. The daily feed cost for different levels of production was calculated based on prevailing prices of the feed components and using different feeding values in a sensitivity analysis. Table 2 shows that it can be cheaper to use the ration that produces 2 x 250 g rather than 1 x 500 g LWG. In that example, the secondary effects on health and reproduction are disregarded as well as the costs for labour and housing because the example is that of a small farm situation; those effects are easily compensated by non-inclusion of value for dung, security, etc.

Practical situations where the highest financial return is not obtained at high individual production levels are well documented, even for so-called developed countries. Compensatory gain in periods of cheap feed combined with underfeeding in periods of feed scarcity lead to (seasonable) adjustment of production levels and calving periods in low input systems of temperate and tropical areas (De Boer and Welsch, 1977; Hermans *et al.*, 1989; Wright *et al.*, 1989). The highest total production per land area can be achieved by reduced individual animal production levels as shown for beef by Jones and Sandland (1974). That supplementation for higher production levels does not always pay is also shown for temperate conditions by Farrington *et al.* (1989). In low input situations with periodic feed scarcity, mere survival can be a realistic goal, requiring feeding objectives, strategies and evaluations that differ considerably from high input situations. This same principle was described by Comella 2000 years ago for conditions in the Roman empire: 'where fodder is scarce, cows should only be allowed to calve every second year' (White, 1970).

A well-known system of dairy farming with economical but reduced and seasonal production is practised in New Zealand. The genetic potential of those cows is similar to that of their close relatives in Europe (Jasiorowski *et al.*, 1987), but the roughage/concentrate/milk price ratio causes production levels to be lower (Bryant, 1986).

Table 1.   The effect of supplementation to a basal ration of medium quality roughage on liveweight gain (LWG), feed conversion ratio (CCR) and cost of feed per kilogram LWG (based on Schiere *et al.*, 1988).

	Coconut cake level[1] (kg day[-1] per animal)	
	0.00	0.75
LWG (g day[-1])	154	322
FCR (total feed DM kg[-1] LGW)	21	12
CCR (concentrate DM kg[-1] LGW)	0	2.34
Feed cost (Sri Lankan Rs kg[-1] LWG)	7.22	12.67

[1] Air dry basis.

Table 2.   Sensitivity analysis on cost of feeding (Sri Lankan Rs day[-1] per animal) for animals at two levels of liveweight gain (A.J. Nell and J.B. Schiere, unpublished data, 1985).

Liveweight gain (g day[-1] per animal)	Assumption[1]		
	I	II	III
250	1.86	1.96	1.86
500	5.00	7.15	3.33

[1] The assumptions relate to different feed costs and variation in feed values/requirements.

In conclusion, variable ratios between the cost of production and the value of the produce result in production optima that are not always most profitable at high production levels. This invalidates the paradigm that high individual productions (of a single commodity) imply a high farm income. As per the resource situation, some systems need to maximise output per animal, other systems need to maximize output per area of land, unit of labour or cash input. The targeting of regional production quota in terms of simple outputs based on extrapolation of current demand as done by Alexandratos (1988), together with aggregation of feed in a gross FCR can lead to inefficient use of resources. On the one hand, high individual production of ruminants based on projected high demand for beef may require concentrate feeds that are more efficiently used in monogastrics. On the other hand, adjusted (lower) individual production levels of ruminants, based on roughages potentially allow a more efficient use of concentrate feeds than are common in monogastric nutrition (Table 1). Whether for high or for low input conditions, the use of feeding standards is essential to understand and predict trends in animal production.

CONCLUSIONS

The usefulness of feeding standards can only be discussed when objectives and the context of their application are clearly defined. If standards are defined as data sets, it can be said that they are used and developed by farmers as well as scientists and policy makers. In principle, the standards of farmers and scientists are complementary, but in practice their formal expression and purpose of application differ considerably. The distinction between tropical and temperate systems is not useful for the discussion of the applicability of feeding standards. An alternative classification is proposed by distinguishing between high and low input systems. In high input systems, the feed is adjusted to the production level of the animal, and in low input systems the production of the animal is adjusted to the feed availability. In most systems a proper mix of the two approaches is required, but the principal point is that high individual production is not always economically attractive. High and low input systems occur under tropical as well in temperate conditions. Further improvement of feeding standards in terms of technical validity of the data sets is required, but that should not conceal that much progress has been achieved over the past decades. The careful use of feeding standards for field application is indispensable since it circumvents the need for a large number of often imprecise and expensive field and station trials. It can help to understand farmers' practices as well as to formulate better research and extension policies.

ACKNOWLEDGEMENTS

Thanks are due to the farmers, scientists and extension workers who helped to develop these ideas, and especially to M.G. Jackson and M. Gill for their comments and suggestions.

REFERENCES

Alexandratos, N. (ed.), 1988. World Agriculture Toward 2000. FAO study, Belhaven, London, 388 pp.
Altona, R.E., 1966. Urea and biuret as protein supplements for range cattle and sheep in Africa. Outlook Agric., 5: 22-27.
Balch, C.C. and Van Es, A.H.J., 1986. Recent advances in feed evaluation and rationing systems for dairy cattle in extensive and intensive production systems. Bull. 196, International Dairy Federation.
Baldwin, R.L. and Hanigan, M.D., 1990. Biological and physiological systems: animal sciences. In: J.W.G. Jones and P.R. Street (eds.), Systems Theory Applied to Agriculture and Food Chain. Elsevier Applied Science, London/New York, 365 pp.

Barker, D.J. and Stoate, J.T., 1969. Survival feeding of cattle during drought. J. Agric. West. Austr., 10(10) October.

Behnke, R.H., 1985. Measuring the benefits of subsistence *versus* commercial livestock production in Africa. Agric. Syst., 16: 109-135.

Birkelo, C.P., Johnson, D.E. and Phetteplace, H.P., 1991. Maintenance requirements of beef cattle as affected by season on different planes of nutrition. J. Anim. Sci., 69: 1214-1222.

Bryant, A.M., 1986. Major determinants of milk production from grazed pasture in New Zealand. Bull. 199, International Dairy Federation, Brussels, pp. 3-6.

Chenost, M. and Reiniger, P. (eds.), 1989. Evaluation of Straws in Ruminant Feeding. Elsevier Applied Science, London, 182 pp.

Chesworth, J.M., McKillop, D.F. and Spriggs, D.J., 1989. Combinations of agro-industrial by-products for use in dairy diets formulated by on farm use of a least cost ration system. In: A.N. Said and B.H. Dzowela (eds.), Overcoming Constraints to the Efficient Utilization of Agricultural By-products as Animal Feed. Proc. 4th Annual Workshop, 20-27 October 1987, Institute of Animal Research Monkon Station, Bamenda, Cameroun. ARNAB, Addis Ababa, Ethiopia, pp. 61-79.

Conway, G.R. and Barbier, E.B., 1990. After the Green Revolution. Sustainable Agriculture for Development. Earthscan Publications, London, 205 pp.

Crampton, E.W. and Harris, L.E., 1968. Applied Animal Nutrition, 2nd Edn. Freeman, San Francisco, CA, 663 pp.

Cronjé, P.B., 1990. Supplementary feeding in ruminants: a physiological approach. S. Afr. J. Anim. Sci., 20: 110-117.

Crotty, R., 1980. Cattle, economics and development. Commonwealth Agricultural Bureaux, Slough, UK. 253 pp.

De Boer, A.J. and Welsch, D.E., 1977. Constraints on cattle and buffalo production in a North Eastern Thai village. In: R.D. Stevens (ed.), Tradition and Dynamics in Small Farm Agriculture. Iowa State University Press, Ames, pp. 115-141.

De Boer, F. and Bickel, H., 1988. Livestock feed resources and feed evaluation in Europe; present situation and future prospects. EAAP Publ. No.37, Elsevier, Amsterdam, 408 pp.

De Wit, J., Dhaka, J.P. and Subba Rao, 1993. Relevance of breeding and management for more or better straw in different farming systems. In: Kiran Singh and J.B. Schiere (eds.), Feeding of Ruminants on Fibrous Crop Residues. Proc. Workshop, 4-8 February 1991, NDRI-Karnal, ICAR, New Delhi, India.

DiConstanzo, A., Meiske, J.C. and Plegge, S.D., 1991. Characterization of energetically efficient and inefficient beef cows. J. Anim. Sci., 69: 1337-1348.

Dirven, J.G.P., 1977. Beef and milk production from cultivated tropical pastures, a comparison with temperate pasture. Stikstof: Dutch Nitrogenous Fertilizer Review, 20: 2-15.

Doyle, C.J., 1974. Productivity, technical change and the peasant producer, a profile of the African cultivator. Food Research Institute Studies, Stanford University, Vol. XIII, No. 1, pp. 61-76.

Doyle, P.T., Devendra, C. and Pearce, G.R., 1986. Rice straw as a feed for ruminants. IDP, Canberra, 117 pp.

Elliot, R.C., Mills, W.R. and Reed, W.D., 1966. Survival feeding of afrikander cows. Rhod. Zamb. Mal. J. Agric. Res., 4: 69-75.

Farrington, T., Marsh, D.K.V. and Mayer, L., 1989. Winter fattening of beef cattle: from research station trials to implementation by farmers. Agric. Syst., 29: 165-178.

Forbes, J.M., 1986. The Voluntary Food Intake of Farm Animals. Butterworth, London, 206 pp.

Frisch, J.E. and Vercoe, J.E., 1978. Utilizing breed differences in growth of cattle in the tropics. World Anim. Rev., 25: 8-12.

Gill, M., 1991. Modelling nutrient supply and utilization by ruminants. In: W. Haresign and D.J.A. Cole (eds.), Recent Advances in Animal Nutrition. Butterworth-Heinemann, pp. 225-236.

Gohl, B., 1981. Tropical feeds. Feed information summaries and nutritive tables. FAO Animal Production and Health Series No. 12. FAO, Rome, 529 pp.

Harris, M., 1965. The myth of the sacred cow. In: A. Leeds and A.P. Vajda (eds.), Man, Culture and Animals, the Role of Animals in Human Ecological Adjustments. Publ.78, AAAS, Washington DC, pp. 217-218.

Hermans, C., Udo, H.M.J. and Dawood, F., 1989. Cattle dynamics and their implications in Pabna district, Bangladesh. Agric. Syst., 29: 371-384.

Ibrahim, M.N.M., 1988. Feeding Tables for Ruminants in Sri Lanka. Animal Feed Advisory Committee, Veterinary Research Institute, Gannoruwa, Sri Lanka.

Jackson, M.G., 1981a. Who needs feeding standards? Anim. Feed Sci. Technol., 6: 101-104.

Jackson, M.G., 1981b. A new livestock development strategy for India. World Anim. Rev., 37: 2-8.

Jasiorowski, H.A., Stolzman, M. and Reklewski, Z., 1987. International FAO black and white cattle strain comparison (1974/84). World Anim. Rev., 62: 2-15.

Jones, R.J. and Sandland, R.L., 1974. The relation between animal gain and stocking rate. J. Agric. Sci., 83: 335-342.

Kearl, L.C., 1982. Nutrient Requirements of Ruminants in Developing Countries. Utah State Institute, Logan, 381 pp.

Kelley, T.G., Parthasarathy Rao, P. and Walker, T.S., 1991. The Relative Value of Cereal Straw Fodder in Semi-arid Tropics of India: Implications for Cereal Breeding Programs at ICRISAT. Paper presented at Conference of Rockefeller Social Science Research Fellows, 2-5 October 1990, Ibadan, Nigeria. Resource Management Program, Economics Group, Progress Report 105.

Ketelaars, J.J.M.H. and Tolkamp, B.J., 1991. Toward a new theory of feed intake regulation in ruminants. Ph.D. Thesis, Agricultural University Wageningen, Wageningen, 253 pp.

Kidane, H., 1984. The economics of farming systems in the different ecological zones of Embu District (Kenya), with special reference to dairy production. Ph.D. Thesis, Hannover University, Hannover, 227 pp.

Kristensen, A.R. and Thysen, I., 1991. Economic value of culling information in the presence and absence of a milk quota. Acta Agric. Scand., 41: 129-135.

Laurenz, J.C., Byers, F.M., Schelling, G.T. and Greene, L.W., 1991. Effect of season on the maintenance requirements of mature beef cows. J. Anim. Sci., 69: 2168-2176.

Leng, R.A., 1990. Factors affecting the utilization of 'poor quality' forages by ruminants particularly under tropical conditions. Nutr. Res. Rev., 3: 277-303.

Mace, R. and Houston, A., 1989. Pastoralist strategies for survival in unpredictable environments: a model of herd composition that maximises household viability. Agric. Syst., 31: 185-204.

McDonald, P., Edwards R.A. and Greenhalgh, J.F.D., 1973. Animal Nutrition. Longman, London/New York, 479 pp.

McDowell, L.R., Conrad, J.H., Thomas, J.E. and Harris, L.E., 1974. Latin American Tables of Feed Composition. Gainesville, FL, 509 pp.

McGraham, N., 1983. Feeding standards, an outmoded concept in ruminant nutrition. In: D.J. Farrell and Pran Vohra (eds.), Recent Advances in Animal Nutrition in Australia. University of New England Publishing Unit, Armidale, pp. 184-191.

National Research Council, 1987. Predicting Feed Intake of Food Producing Animals. National Academy Press, Washington, DC, 85 pp.

Nell, A.J., Schiere, J.B. and Ibrahim, M.N.M., 1986. Economic evaluation of urea ammonia treated straw as cattle feed. In: M.N.M. Ibrahim and J.B. Schiere (eds.), Rice Straw and Related Feeds in Ruminant Rations. Proc. Int. Workshop, 24-28 March 1986, Straw Utilization Project, Kandy, Sri Lanka, pp. 164-170.

Nordblom, T. and Halimeh, H., 1982. Lentil crop residues make a difference. ICARDA/University of Saskatchewan, Canada, LENS Newsl. No. 9, pp. 8-10.

Oddy, H., 1978. Feed requirements of sheep and cattle during drought using a metabolizable energy system. AGBull. 3, Department of Agriculture, New South Wales, Australia, 22 pp.

O'Donovan, P.B., 1984. Compensatory gain in cattle and sheep. Nutr. Abstr. Rev. Ser. B, 54(8): 389-410.

Ørskov, E.R., Ried, G.W. and Kay, M., 1988. Prediction of intake by cattle from degradation characteristics of roughages. Min. Prod., 46(1): 29-34.

Oscarsson, G., 1975. The future of animal production: economic aspects. In: R.L. Reid (ed.), Proc. Third World Conference on Animal Production, 22-30 May 1973, Sydney, University Press, Thema 5, Pap. 57, pp. 447-457.

Owens, F.N. (ed.), 1987. Feed intake by beef cattle. Proc. Workshop, 20-22 November 1986, Oklahoma State University, Stillwater, OK, 399 pp.

Patel, R.K., Kumar, P., Gangadharan, T.P., Dhaka, J.P., Voegele, K., Sukumaran Nair, R. and Sreekumaran Nair, G., 1977. Economics of Crossbred Cattle. A study of the Cattle Breeding Programme of Indo-Swiss Project, Kerala. NDRI-Karnal, India.

Potts, G.R., 1982. Application of research results on by-product utilization: economic aspects to be considered. In: B. Kiflewahid, G.R. Potts and R.M. Drysdale (eds.), By-product Utilization for Animal Production. Proc. Workshop on Applied Research, Nairobi, Kenya, 26-30 September 1982, IDRC, Ottawa, Ont., pp. 116-127.

Prasad, C.S., Sampath, K.T., Shivaramaiah, M.T. and Walli, T.K., 1993. Evaluation of cereal and millet straws for ruminant feeding based on *in-vivo* studies - a review. In: Kiran Singh and J.B. Schiere (eds.), Feeding of Ruminants on Fibrous Crop Residues. Proc. Int. Workshop, 4-8 February 1991, NDRI-Karnal, ICAR, New Delhi, India.

Preston, T.R., 1985. Validity of feeding standards and development of feeding systems based on crop residues and agro-industrial by-products. In: T.R. Preston, Vappu L. Kossila, J. Goodwin and S.B. Reed (eds.), Better Utilization of Crop Residues and By-products in Animal Feeding: Research Guidelines. 1. State of Knowledge. FAO Anim. Prod. Health Pap. 50, FAO, Rome, pp. 197-213.

Rangnekar, D.V., 1993. Farmers' perceptions of quality and value of feeds, fodder and feeding systems. p. 415-422 in Kiran Singh and Schiere, J.B. (eds), Feeding of Ruminants on Fibrous Crop Residues, proc. of a workshop held from Feb 4-8, 1991 at the NDRI, Karnal, publ. by the Indo Dutch BIOCON Project, Indian Council of Agricultural Research, Krishi Bhavan, Delhi, India.

Ranjhan, S.K., 1990. Nutritional Value of Animal Feeds and Feeding of Animals. ICAR, New Delhi, India.

Ranjhan, S.K., 1991. Chemical Composition and Nutritive Value of Indian Feeds and Feeding of Farm Animals. ICAR, Krishi Anusandhan Bhaven, New Delhi, 132 pp.

Rayburn, E.B. and Fox, D.G., 1990. Predicting growth and performance of Holstein steers. J. Anim. Sci., 68: 788-798.

Robinson, J.J., 1990. Nutrition in the reproduction of farm animals. Nutr. Res. Rev., 3: 253-276.

Schiere, J.B. and De Wit, J., 1995. Feeding of urea ammonia treated straw in the tropics (II): Assumptions on nutritive value and their validity for least cost ration formulation. Anim. Feed Sci. Technol., 51: 45-63.

Schiere, J.B. and Nell, A.J., 1993. Feeding of urea ammonia treated straw in the tropics (I): A review of its technical principles and economics. Anim. Feed Sci. Technol., 43: 135-147.

Shah, T., Tripathi, A.K. and Desai, M., 1980. Impact of increased dairy productivity on farmers' use of feedstuffs. Economic and Political Weekly, India, August 16, pp. 1407-1412.

Shanti George, 1991. The female of the species: women and dairying in India. In: H. Singer and Prendergast (eds.), Development Perspectives in the 1990's. MacMillan, UK, in press.

Wahed, R.A., Owen, E., Neate, M. and Hosking, B.J., 1990. Feeding straw to small ruminants: the effect of amount offered on intake and selection of barley straw by goats and sheep. Anim. Prod., 52: 283-289.

Welsch, D.E., 1975. Resource use in systems of intensive animal production. In: R.L. Reid (ed.), Proceedings of the Third World Conference on Animal Production. Sydney, University Press, Thema 4, Pap. 52, pp. 409-415.

White, K.D., 1970. Roman Farming. Thames and Hudson, London.

Wright, I.A., Russel, A.J.F. and Hunter, E.A., 1989. Compensatory growth in cattle grazing different vegetation types. Anim. Prod. 48: 43-50.

Zemmelink, G., 1980. Effect of selective consumption on voluntary intake and digestibility of tropical forages. Agric. Res. Rep. 896, PUDOC, Wageningen, 100 pp.

Zemmelink, G., Brouwer, B.O. and Ifar, S., 1992. Feed utilization and the role of ruminants in farming systems. In: Hari Purnomo, M.N.M. Ibrahim, R. De Jong and J. Van Bruchem (eds.), Proc. Workshop Livestock and Feed Development in the Tropics. 21-25 Oktober 1991, Malang, Indonesia. Agricultural University Wageningen, Wageningen and the Commission of the European Communities, Brussels.

MATCHING ANIMALS AND FEEDS
FOR MAXIMUM FARM SYSTEM OUTPUT IN LOW INPUT AGRICULTURE; EXPLORATORY THOUGHT EXPERIMENTS[1].

J.B.Schiere, J. De Wit and F.A. Steenstra

SUMMARY

Changing resource / demand patterns in agriculture require the design of new farming systems. Thought experiments can serve as a form of such New Farm System Development (NFSD), and they are used here to match livestock and feed supplies for maximum output of closed systems. Sensitivity analysis is done with linear programming (LP) to simulate system behaviour with varying feed qualities and animal production levels. Milk yield and animal numbers are measures of system output in two hypothetical and simplified cases that represent actual farming systems. The level of animal production, including maintenance itself, ranges from 0.75 to 3.00 times maintenance in both cases. The first case considers the total feed as one aggregate, with a nutritive value ranging from that of straw to good quality forage. It establishes the type of animal that needs to be used to achieve maximum system output, i.e. the individual animal output is matched to the feed supply. The second case allows animals with different production levels to select between two feeds that are mixed in different proportions, but representing the same feed quality scale as in the first case. This latter approach allows the adjustment of the animals to feed as well as adjustment of feed to the animals for maximum system output. The results of both cases confirm that better feed and higher individual animal output tend to increase total system output in terms of milk by reducing the number of animals. Nevertheless, the term 'damning objective' is introduced to express that high individual animal output targets can reduce the total system output if they exceed the system resources. Desaggregation of feed resources, i.e. the possibility of selective consumption, increases the system output in terms of milk, sometimes even by increasing the number of animals with low individual output. An important social trade-off of higher system output in terms of milk with fewer animals, by increasing feed quality, is the possibly uneven distribution of animals among farm systems. The results agree with farmers' practice and situations reported in the literature. Issues such as the damning objective and the need for desaggregation of feed imply that traditional, additive feed balances with fixed and average animal production targets are likely to misinterpret system behaviour. Issues for further research and biases due to modelling artifacts are discussed.

---

[1] This chapter is based on a presentation "Optimization of Natural Resources for Sustainable Animal Production" at the First International Animal Nutrition Workers Conference for Asia and the Pacific, Sept. 23-28, 1991, organized by the Animal Nutrition Society of India in Bangalore, India.

INTRODUCTION

Changing resource / demand patterns, force both low and high input crop and/or livestock systems to reassess their methods and objectives of production (Crotty, 1980; Hayami and Ruttan, 1985; Ch. 2.2 and 2.3). Erosion, overgrazing, and soil mining are common problems in low external input agriculture (LEIA). When fibrous crop residues are available, the improvement of straw quality/quantity plays a role by adjusting cropping practices (Patil *et al.*, 1993, Joshi *et al.*, 1994). In systems with more access to resources, problems of low nutritive value of straw can be overcome by use of supplements, or physical and/or chemical treatments (Sundstol and Owen 1984; Kiran Singh and Schiere, 1993; Ch. 4.1). In high external input agriculture (HEIA), a major physical limitation of the production system lies in the disposal of excess minerals from specialized animal production (Durning and Brough, 1991; Kaasschieter *et al.*, 1992a). Also, straw disposal becomes a problem for specialized crop farmers, e.g. in the grain belt of eastern UK (Staniforth, 1982; Kelley, 1992) as well as in many tropical farming systems where straw is burned. The build-up of plant pests and disease, as well as wind or water erosion due to mono cropping, can make it attractive to include a ley or a catch crop, also to reduce dependence on agrochemicals (Johnston, 1972; Gibson, 1987; Woodward and Foster, 1988).

Adjustment or design of new farming systems, based on changing resource / demand patterns, is part of what Simmonds (1986) calls New Farming Systems Development (NFSD). One form of NFSD is the use of 'thought experiments' as applied by Von Thünen and contemporaries at least 150 years ago, which constituted a form of modelling to understand system behaviour (Nou, 1967; Ch. 2.1). This paper uses thought experiments to explore the design of systems that maximise animal output from a given quantity and varying quality of feed at a wide range of individual animal production levels. The paper first discusses the terminology and methodology related to the modelling employed. It then explores the central question, i.e. the adjustment of animals and feeds for maximum system output.

By starting from a situation in a low input system this paper assumes that livestock depend on the supply of feed biomass from the crop subsystem: a variable mix of poor quality fibrous crop residues or high quality fodders and brans. Purchase of feed is not allowed in our cases; the calculations explore possibilities for animal production without use of external feeds. The thought experiments focus on the behaviour of individual farm systems at farm level, but they also explain mechanisms that are relevant for the village, regional or even higher level. The first case concerns the adjustment of animals to the feed, the second case concerns the mutual adjustment of feed and animals (Ch. 2.2 and 5.1).

In order to see the wood for the trees, system behaviour, in this chapter, was simulated with highly simplified models, measuring only milk yield and number of animals to represent outputs such as meat, offspring, dung, draught or security. An economic assessment was not attempted because the thought experiments aimed to understand feed allocation patterns on an abstract level. The focus on feeds is justified by the important role of feeds as a source of energy in livestock systems: biologically speaking, animal production is essentially the conversion of feed energy into animal products, and the resultant energy flow is a major determinant of system behaviour (Odum, 1983; Ch. 2.2 and 6).

MATERIALS AND METHODS

Thought experiments can be done by either using mental arithmetic, a stick in the sand of the farmyard, or with powerful computers. We have chosen for simple calculations with personal computers and existing software.

Models and software

Several modelling approaches and software packages are available for feed allocation and simulation of livestock systems in general (Zemmelink *et al.*, 1992; Udo and Brouwer, 1993). Linear programming (LP) was used here because it is specifically designed for resource allocation. It also provides a convenient platform for interdisciplinary discussion, and a range of software is available. Many, if not all, drawbacks of LP can be accounted for, such as the assumption of additivity and linearity, besides rigid decision-making (Romero and Rehman, 1989; Van Niejenhuis and Renkema, 1989). Resource allocation over time and space can be done by LP with no difficulty other than an expanding matrix size. LP is often understood to give one solution rather than a range, but this issue can be overcome by running the model several times (Renkema, 1972; Morrison *et al.*, 1986; Kingwell and Pannell, 1987; Ch. 4.1). Indirectly, this approach also allows the use of a smaller matrix by reducing the number of variables, as done in these cases. The repeated runs and the recording of outputs for the thought experiments of this paper were automated with a set of macros in LOTUS 1-2-3, version 2.0 (registered trademark of LOTUS Development Corporation).

Testing and use of models

Testing of results from thought experiments is difficult since such experiments are precisely intended to understand problems beyond practical experimentation. Moreover, discrepancies between the model and practice can originate not only from imperfections in the data and relations of the model, but also from suboptimal farmers' practice (Sol *et al.*, 1984; Morrison *et al.*, 1986). Model solutions should be tested nevertheless, by one or more of the following approaches:
- after predicting the behaviour of a particular system it should be possible to predict its occurrance under practical conditions. Essentially this is a form of deductive reasoning, a common approach in physics and astronomy that can also be used to predict feasibility of straw feeding methods (De Wit *et al.*, 1993; Ch. 4.1).
- modelling can be tested by checking underlying calculations (Ch. 4.2), or by testing the results against general laws and analogies from e.g. physiology, economics or thermodynamics (Ch. 6).
- a model can also be tested by inserting extreme values.

All three approaches are used in this paper, even though testing and use of the models are often difficult to distinguish.

System output and animal units.

The system output was measured as milk yield (4% fat-corrected milk), numbers of animals, and feed used or refused. No allowance was made for calf crop, milk consumed by the calf, herd components such as bulls or growing animals, nor for meat, draught or dung production. The advantages and biases of this simplification are explained in the discussion. Animal subsystem output was expressed as multiples of an animal unit for maintenance, based on TDN (AUM_{TDN}), calculated for a ruminant of 350 kg BW (Table 1), called M for ease of notation. The AUM can be refined for protein requirements as also shown in Table 1, but that would not serve the point of our exploration of system behaviour in this paper. An animal of 0.75 M is included here, because survival, even at negative weight gain, is an essential form of animal output in farming systems with fluctuating feed supplies (Allden, 1970: Ch. 5.1).

Table 1   A definition of animal units based on level of production, expressed as multiples of maintenance for requirements for protein (AUM_{CP}) and energy (AUM_{TDN})

AUM		0.75	1.00	1.25	1.50	1.75	2.00	2.25	2.50	2.75	3.00
Milk yield[1]	$(lts.an^{-1}.day^{-1})^{-1})))$0.0001		0.01	2.19	4.39	6.58	8.77	10.96	13.16	15.35	17.54
CP-Maint.	$(kg.an^{-1}.day^{-1})^{-1})))$0.29		0.29	0.29	0.29	0.29	0.29	0.29	0.29	0.29	0.29
CP-Milk	$(kg.an^{-1}.day^{-1})$	-0.20	0.00	0.20	0.40	0.60	0.79	0.99	1.18	1.38	1.58
CP-Total[2]	$(kg.an^{-1}.day^{-1})$	0.09	0.29	0.49	0.68	0.88	1.08	1.27	1.47	1.67	1.87
TDN-Main-Mt	$(kg.an^{-1}.day^{-1})$	2.82	2.82	2.82	2.82	2.82	2.82	2.82	2.82	2.82	2.82
TDN-Milk	$(kg.an^{-1}.day^{-1})$	-0.71	0.00	0.71	1.41	2.12	2.82	3.53	4.24	4.94	5.65
TDN-Total[2]	$(kg.an^{-1}.day^{-1})$	2.12	2.82	3.53	4.24	4.94	5.65	6.35	7.06	7.77	8.47
AUM_{CP}		0.32	1.00	1.69	2.37	3.06	3.75	4.44	5.12	5.81	6.50
AUM_{TDN}	(M)	0.75	1.00	1.25	1.50	1.75	2.00	2.25	2.50	2.75	3.00
F_P		0.88	1.00	1.12	1.25	1.38	1.50	1.62	1.75	1.88	2.00

[1]   This table gives a milk yield for AUM 0.75 and 1.00 as used in the objective function of the LP matrix; theoretical milk output should be resp. -2.19 and 0.0 $lts.an^{-1}.day^{-1}$ (see text).

[2]   Requirements are based on NRC(1988; Table 6.3); 3.55 g $CP.kg^{-0.75}$ and 34.9 g $TDN.kg^{-0.75}$ for maintenance, and 90 g CP with 322 g TDN per liter of milk.

The cases

These thought experiments were applied to two imaginary cases that reflect field conditions, and a sensitivity analysis of coefficients and relations was done in both cases. The important difference between case I and II is that in the first, feed is one homogenous inseparable mix of bad and good feed, selection between the feed components is not possible. The second case however, uses a separable mix of two feeds, good and bad, permitting selection between feed components by the farmer and/or animal. The average nutritive value of the available feed is the same over the horizontal axis from left to right, in case I and II (see Figures 1 and 2). Due to the possibility of selection between feeds however, the nutritive value of the intake from the mix at any given point of the X-axis, may differ between case I and II, as chosen by the model to achieve highest system output. Nutritive values expressed as TDN40/CP4, indicate that the value of total digestible nutrients (TDN) is 40 and the crude protein content (CP) is 4, expressed as percentage of dry matter. The effect of changes in quality of feed and type of animal are predicted for:

CASE I: one homogenous feed, consisting of an inseparable mix of good and bad feed, selection within feeds is not possible:

Ia: a fixed quantity of feed, representing a mix of two feeds, e.g. straw (TDN40/CP4) on the one hand, and an excellent fodder (TDN70/CP16) on the other extreme. The ratio of their mix changes from 100/0 to 0/100 on the X-axis of the figures, with a corresponding improvement of nutritive value. The feed is offered to animals ranging from 0.75*M to 3.00*M in increments of 0.25*M, whereby intake is allowed to increase with the individual output. This is done by introducing F_p a factor that corrects the DMI according to the level of milk production expressed per multiple of maintenance (ARC, 1980; B.J. Tolkamp, pers.comm., 1994), see the bottom row of Table 1. The value F_p starts at 1 for animals at 1*M and increases linearly to 2 for animals at 3*M, being 0.875 at 0.75M.

Ib: as for Ia, except that F_p remains constant (i.e. one) for all cow production levels.

Ic: as for Ia, except that the feed value on the X-axis runs from TDN55/CP10 to TDN65/CP14 with smaller increments.

CASE II: the feed offered consists of two feeds, again good and bad, but selection within feeds is possible:

IIa: same as case Ia, but the feeds are desaggregated to allow rejection of feed i.e. selective consumption; F_p ranges from 1-2.

IIb: same as IIa, but the basal feed is of better quality, e.g. 'untreated' straw is replaced with 'treated' straw of TDN55/CP10,

IIc: same as IIb, but the poor quality feed is further 'improved' to TDN65/CP14, approaching the quality of good grass.

IId: same as IIb, but F_p remains 1 for all levels of animal production.

The matrix

Only one small matrix is required for the cases in this paper (Table 2), because different variables are tried over repeated runs, rather than in one single run. The coefficients differ per case and they are indicated above, and in the figures with the results. The matrix is explained as follows:

- *objective values*

* OF_i: cost of feed, here valued at '0' in all cases. The use of the feeds is restricted only by their availability and nutritive value (see constraint CF_i).

* OS_i: cost or value of feed not fed (VS_i), is also '0', being a so-called store value, as further explained under 'variables'.

* OA_i: animal output measured as milk production (l/animal/day). The objective values are .001 for 0.75M, .01 for 1M, 4.4 for 1.5M, proceeding with constant increments via 2M (8.8 lts); 2.5M (13.2 lts) to 3M (17.5 lts). These values are taken from on Table 1 where 8.77 litres of milk equals 1M, in the sense that a cow with a milk output of 1*M, has a real production of 2*M, i.e. maintenance + 1 * maintenance in terms of milk. The use of .0001 and .01 in the objective row for animals at 0.75 and 1M ensures that the

model picks up animals at sub maintenance and maintenance, with values that can easily
be recognized and that do not inflate the total objective value.

- *variables*
    * the sum of VS_i + VF_i is the total feed offered, VF_i is feed consumed and VS_i is a 'store'
    for refused feed. VS_i allows animals to refuse feed, e.g. when the DMI of a feed is too
    low to satisfy nutrient requirements. The VS_i can be deleted by making VF_i the
    maximum amount of feed to be fed, its slack value then represents VS_i. That would
    however, complicate the use of VS_i for other applications e.g. for transfers to other
    seasons or for use as bedding, mulch or thatching.
    * VA_i is the number of animals at a given level of production that are used in the model.

- *constraints*
    * CF_1 is the yield of good feed, e.g. a legume or young grass, estimated to be 7200 kg
    DM ha^{-1} year^{-1} (= 20 kg day^{-1}). CF_2 is the yield of poor feed such as straw from either
    wheat or rice from a total area of 1 ha. Its maximum value is 7200 kg ha^{-1}, i.e. 20 kg day^{-1}
    based on two grain harvests per year of 3000 kg ha^{-1} yield^{-1}, and not accounting for the
    yield of brans and ratoon (Insiani 1990). The yields of poor and good feed were assumed
    to be equal, to avoid confounding effects of quality and quantity. For the same reason,
    the organic matter content of all feeds was assumed to be equal, a simplification that did
    not alter the point of this paper.
    * $Cmax_i$ was the maximum DMI values of the feeds, estimated by:
    $$OMI = -42.8 + 2.3039 \times OMD - 0.0175 \times OMD^2 - 1.8872 \times N^2 + 0.2242 \times N \times OMD$$
    (Table 1 of Ketelaars and Tolkamp, 1992a) and
    $$DMI = OMI \times F_a \times OM^{-1} \times F_p$$
    where:
    OMI: organic matter intake, (g kg$^{-0.75}$ d^{-1})
    DMI: dry matter intake, (g kg$^{-0.75}$ d^{-1})
    F_a: animal factor,
    F_p: correction factor for DMI according to animal production level
    * $Ctdn_i$ and Ccp_i represent the rows with the nutrient requirements of the animals, and
    the nutritive values of the feeds used. As described earlier, nutrient requirements for
    maintenance and milk yield were calculated based on NRC (1988).

Table 2   The LP matrix used for all the cases.

	Variable	VF1	VS1	VF2	VS2	VA1	> = <	RHS
	Objective code	OF1	OS1	OF2	OS2	OA1		
	Objective value	0	0	0	0	2.19		MAX
CF1	DM avail (good)	1.00	1.00				=	16.00
CF2	DM avail (poor)			1.00	1.00		=	4.00
Ctdn1	TDN min	0.70		0.55		-3.53	>	0.00
Ccp1	CP min	0.16		0.10		-0.49	>	0.00
Cmax1	DMI max (good)	1.00		1.00		-8.77	<	0.00
Cmax2	DMI max (poor)			1.00		-6.89	<	0.00

note: the coefficients in this case belong to case IIb with a milk production of 1.25*M, i,e. 2.19 ltr.

RESULTS

The main results of the calculations are summarized in Figures 1 and 2. The figures have been given numbers according to the cases, to enable easier referencing. For example, nr 1Ap refers to the production of milk per system for case 1A; 1Aa refers to the number of animals and 2Af refers to the amount of poor feed consumed in case 2A. Because the general features of the figures are similar, the text refers to lines such as MM', F'F'' etc. as in Figure 1 (case IA), to be related with similar lines in the other figures.

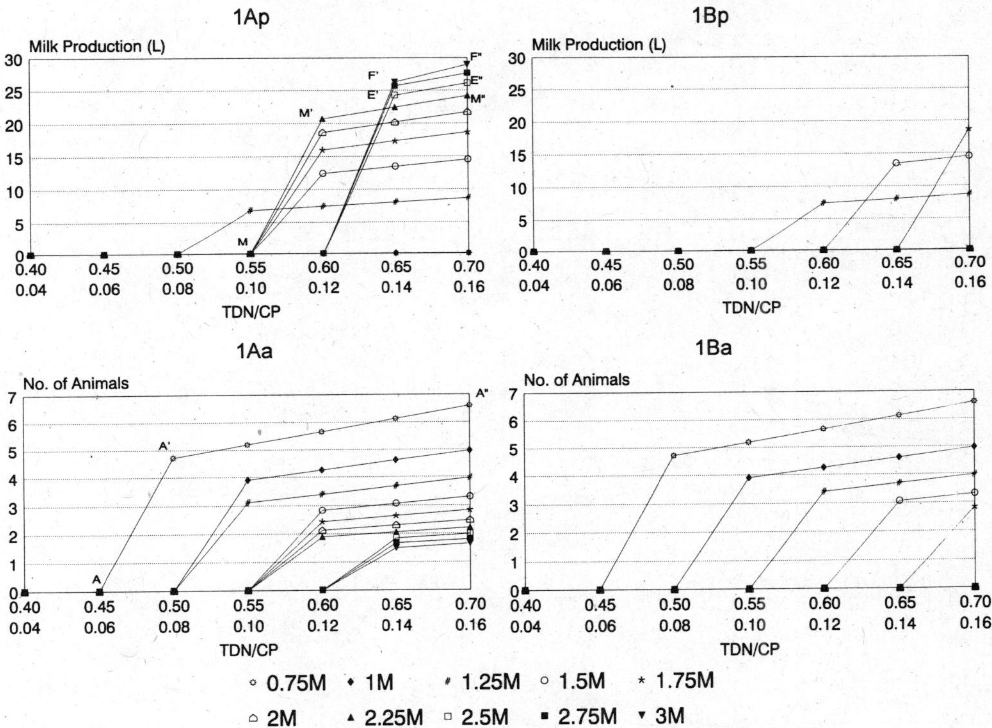

Figure 1:   The effect of changing feed quality and individual animal output on total milk production and number of animals, with (1A) and without (1B) adjusting feed intake for production level of the animal through F_p.

Animal System Output

Better feed of constant quantity, leads in all cases to higher total system output, expressed as milk per system. The increase is achieved, however, mainly by using fewer animals with higher individual production, thus saving on maintenance requirements. When animals of constant individual production are used, better feed permits the system to maintain more animals (line A'A" in Figure 1Aa). However, provided that the feed is good enough, animals with a higher individual output would increase total system output in terms of milk, to a greater extent than the use of more animals of the same output (e.g. F'F'' is preferred over E'EE'' in Figure 1 Ap).

A target for output per animal that exceeds the potential of the available feed quality, reduces the total system output, and this constitutes a principle that we propose to call the

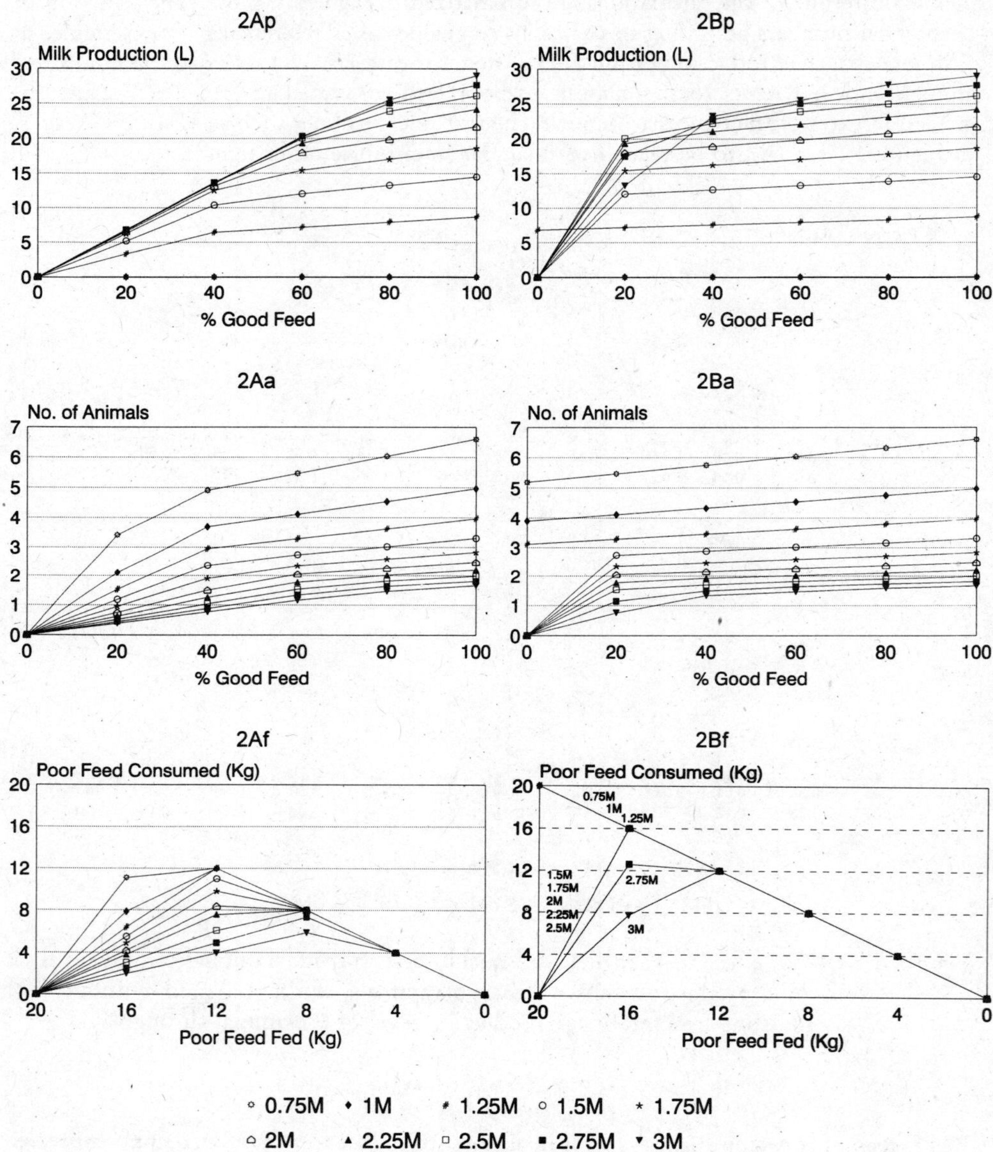

Figure 2:   The effect of a changing mix of feed, in which selection is possible, from feeds with respectively TDN40/CP4 (Figure 2A), or TDN50/CP10 (Figure 2B) to TDN70/CP16, at different levels of individual animal output on total milk output, number of animals, and quantity of feed refused.

damning objective[2], because such a production target cannot be achieved with the resources available within the farm system. The farm system has to choose then either to import feed from other systems, or to reduce the expected performance per animal. In fact, in the case of a damning objective, the on-farm resources cannot even be used if they are not good enough. They become a waste to be disposed of, if no other uses can be found. Depending on the rigidity of the target, the total system output is lower when excessive individual output targets are used. For example, and in extreme terms, at TDN55/CP10 in Figure 1 A_p (case IA), the total system output is higher with 1.25M than with 2, 2.5 or 3M cows. In other words, a lower individual animal output target, results in a higher total system output (in terms of milk), though it is a somewhat artificial result due to the simplification that a 2.50M cow cannot function as a 2.25M cow. The principle is best illustrated in Figure 3 where only at feed compositions of 40% or more good feed, the total system output continues to increase with higher individual animal output. The damning objective does not imply that the animal cannot produce at a lower level, rather it implies that the target should be flexible. The principle and magnitude of the damming objective depend on the quality of the basal feeds (case IIa vs IIb), and on the possibility of selective consumption (case Ia vs IIa).

Desaggregation of feed, as allowed in case II, introduces the possibility of **selective consumption**. It increases total system output in terms of milk (case Ia vs IIa) if other model parameters remain the same. It even allows higher total milk output at a higher total number of animals, particularily at the lower range of feed quality (Figure 1Aa vs 2Aa). The amount of feed refused is shown in Figures 2Af/2Bf, feed refusals being higher at lower qualities of the basal feed and at higher levels of individual animal output.

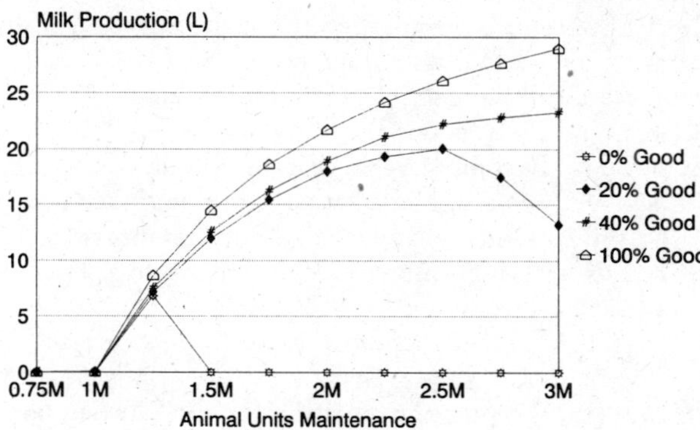

Figure 3:   The effect of feed quality, expressed as percentage good feed offered, and individual animal output on total system output in terms of milk production (based on case IIb)

---

[2] The term 'damning objective' may need to be replaced with another, less ethical sounding, or already existing terminology; but while consulting collegues from other disciplines, no suitable alternative was found until the date of finalization of the manuscript

The effect of improved 'straw' quality is clear from the comparison between Figures 2A and 2B, as well as by proceeding from left to right on the X-axis in all cases. Higher quality of the basal feed, affects the magnitude of the damning objective, and as with desaggregation, it also allows an higher system output by keeping more animals, as less feed is refused (case IIa vs. IIb).

DISCUSSION

The thought experiments of this paper predict the behaviour of closed, i.e. low input animal production systems, by matching feeds of different qualities with animals of different milk yields. The cases employ imaginary feeds and animals, but they reflect actual situations in a variety of LEIA farming systems. The feeds can be understood to represent poor quality roughages such as straw at the left of the X-axis of Figures 1 and 2. They also represent urea treated straw, grass, tree leaves or concentrate supplements, as one proceeds to the right on the X-axis in case I, and an increased ratio of good *vs.* poor feed in case II. Feed quantity is kept constant in our calculations; only the quality is changed.

As was argued previously, the models can be tested by comparing the results with existing situations from the literature or practice, and by inserting extreme values in a sensitivity analysis. The testing by analogy with other system behaviour is done in Ch. 6. Here we have used the first two approaches, and fortunately the DMI predictions are in agreement with those reviewed in Ch. 4.2. Also, the results of the calculations agree with common sense and practice as they predict:
- a generally higher system output in terms of milk, with increased feed quality and individual animal milk yield,
- a positive effect of selective consumption, i.e. desaggregation of feed supply, in animal output in terms of milk and total animal numbers,
- the strategy to use more animals of lower individual milk output when feed quality decreases (Breman and De Wit, 1983)
- when herd survival is given more value than only milk output, the model prefers more animals, with less feed refused, a result that agrees with the conclusion of Zemmelink *et al.* (1992). At the extreme, and of course only temporarily, the 0.75M animals are preferred to animals with high individual production (e.g. case IAa)

Changes in feed quality and animal output

Breman and De Wit (1983) reported a situation for sub-Sahelian zones where feed quality declines. Even though feed quality in their study was confounded by increased quantity of feed, their observations fit the system behaviour in our thought experiments. Moreover those authors identified selective consumption as a farmers' strategy, also described by Zemmelink (1980), Wahed *et al.* (1990) and McFarland *et al.* (1992).

The increased system output achieved with fewer animals of higher individual output is the typical approach followed in HEIA farming systems which solve a shortage of quality or quantity of feedstuffs, by purchase from outside the system. In conditions of extensive grazing, the total output is increased by use of expanded grazing areas and selective consumption by the animals. To benefit fully from improved feed supply however, it is also necessary to increase production factors such as housing, and veterinary care. Here, we assumed that these factors were not limiting, but the problem reflects that a proper balance of all production factors improves total system efficiency (De Wit, 1992). This argument however, also runs the other way: if feed and housing are not adequate, there is no point in having high potential animals. Contrary to a common (HEIA) approach, if the total feed pool comprises of a large proportion of poor quality relative to high quality feeds, the total system output in LEIA could be greatest with many low-producing animals. If this were true, it should theoretically be possible to predict the type and number of animals which should be kept for a given resource supply. It ultimately affects the decision whether to choose cross-breds, locals or high-yielding purebreds. Many livestock improvement programs which failed, gave evidence that the introduction of purebreds led to no, or lower production than the use of adapted animals (DGIS, 1987).

Adjustment of individual animal output to lower quality feed resources, receives less attention than breeding and feeding for higher individual animal output. The work of Frisch and Vercoe (1978) on genotype * environment interactions, and the work by Hayman (1974) and Alexander *et al.*, (1984a,b) on the development of the Australian milking zebu and Friesian Sahiwals, hardly mention breeding for feed utilization. Practical examples of breeding for adjustment to feed supply are nevertheless available, e.g. in the stratification of sheep breeds, such as in the Scottish highlands (Frazer, 1949; Robertson, 1983). There, the breeding objectives were determined by the type of wool, the shape of the animal and disease stress, but the effect of nutrition was explicitly recognized. Frazer (1949, p.147) touches on the principle of the damming objective as he observes for those typically low input conditions:

> *'it is quite **impossible** to produce a first-quality lamb off a barren hill-side. All that the land's fertility will support is the slow growth and slow reproduction rate of hill breeds of sheep [...]. Thus, by a judicious system of [...] crosses the final result, on good lowland pasture, is a combination of the hill breeds' constitution, the Border Leicester's fertility, the Down sheeps' mutton, expressed in the form of twin lambs of Down type drawing abundant milk from a mother of hill descent'.*

The higher system output in terms of milk at better feed quality, is achieved primarily because animals can eat more and better feed (Forbes, 1986; Ketelaars and Tolkamp, 1992), thus reducing the relative amount of feed used for maintenance. An exception is the slope as presented in line M'M'' where animal requirements are met by reducing feed intake as feed quality improves. This situation can also be found in practice, for example where good feed is diluted with poor quality feed, e.g. by chopping, or where animals are given restricted access to feed to avoid overfeeding and to maintain more animals. These options are particularly relevant where demand for dung, moderate draught output, or for savings / investments, make that higher animal numbers are more important than increased individual output of milk, or where feed quality exceeds the animal's genetic capacity for milk output (Zemmelink *et al.*, 1992; De Wit *et al.*, 1993).

Selective consumption

Desaggregation of feeds allows selective consumption (Zemmelink, 1980; 1986; Wahed *et al.*, 1990; McFarland *et al.*, 1992; Zemmelink *et al.*, 1992). The need for selective consumption depends on the desired level of animal production and on the feed availability. As said above, prevention of selective consumption, e.g. by chopping, is only useful if the system prefers to maintain more animals at lower individual milk output, combined with higher preference for dung, draught, and saving account functions (Zemmelink *et al.* 1992; De Wit *et al.* 1993). In practice, this takes place where a specified number of animals are required to pull the plough, or where animals have to survive a lean season to take advantage of cheap liveweight gains in the lush season. In all cases where selective consumption is allowed, the total system output increases in terms of milk and number of animals, though the effect is less pronounced at higher feed qualities. The principle is illustrated by the difference between cases I and II, and it applied by farmers that prefer to burn straw rather than to feed it to their animals (Staniforth, 1982; Kelley, 1992). Effects of straw on rumen function as the prevention of bloating or acidosis are disregarded here, but the model can be expanded to include requirements for minimum levels of roughage if necessary.

The damning objective

The damning objective means that nutrient requirements for animals with individual output targets which are beyond the resources of the farm system, can only be met by importing feed from outside, i.e. by 'externalizing a shortage'. It also implies that feeds from within the system cannot be used, resulting in the burning of straw or 'externalization of a waste', on specialized crop farms of HEIA in Western Europe where farmers cannot keep animals suitable for the use of straw. Due to their high and rigid production targets, those animals cannot assist with straw disposal (Staniforth, 1982; Kelley, 1992). This is in contrast to the situation of farmers in LEIA systems, who value straw as a maintenance feed, to keep animals through a lean season (Insiani, 1990; Ifar *et al.*, 1995). Even farmers in high input systems occasionally feed straw in winter, to take advantage of cheaper good feed in the lush season, i.e. they accept a reduction of the individual output target in order to better utilize on-farm resources.

Two crucial issues arise from the damning objective in relation to general system behaviour, particularly as the individual animal can be imagined to represent the total animal- or even farm subsystem. The first point is, that if the output target of an (animal) subsystem is too high, it is likely to negatively affect the output from the crop or adjoining animal subsystems, and hence of the overall system at a higher level in the hierarchy. Following this principle, not much imagination is required to see that, the sustainability of one subsystem cannot be established without taking into account the sustainability of the overall system. Sustainability of (livestock, crop, etc.) production, therefore needs to be considered within the framework of total system sustainability at a higher level in the hierarchy. This problem is inherently linked with the problem and danger of defining system boundaries, i.e. the tendency towards reductionism that causes problems of internalization and externalization (Conway and Barbier, 1990; Daly and Cobb, 1990; Ch. 2.1).

The second point is that with the chosen hypothetical, but realistic, feeds, the effect of the damning objective seems to be most pronounced where crop residues form a relatively large part of the total feed. There are indications that most added value of livestock in mixed crop-livestock systems, biologically speaking, takes place at lower feed qualities. It is precisely at that end, where only a fraction of the land is used for feed production, either on-farm or on roadsides and wastelands (Kaasschieter *et al.*, 1992b; Schiere, 1992). The estimates will vary widely between systems, but the logic is appealing. In systems where cropping is possible, animals produce fewer nutrients per area unit than crops (Spedding, 1987). Consequently, a sizeable animal component in mixed systems based on fodder production at the expense of crops, would negatively affect the total system output in terms of food produced for humans. If this is true, it implies that in LEIA, animals add value particularly when they can use crop residues at adjusted animal production levels. If cash enters the system, for example by sale of milk, the farmers either replace grain and straw production by specialized fodder, or they start to buy supplements. The systems become open and they change from LEIA to HEIA characteristics, towards systems with better feeds and higher milk output per animal, at the expense of food production in terms of calories and protein for human consumption.

Total system output and equity

In terms of equity, i.e. the distribution of control over production and resources, these thought experiments also provide interesting points for further study. This stems from the fact that the reasoning starts from 'closed' system conditions, where due to distribution problems, resource shortage cannot be masked by using external inputs. In the first place, it is necessary to recognize that inputs originate from somewhere, whether from common property such as communal grazing or fossil energy, or from other systems (Ch. 6). The extraction of feed resources from the weaker system[3], to meet the damning objective of the stronger system, will increase the total system output only if those resources could not otherwise be used in the weaker system. For example, the production of the crossbred cow can increase if the straw - after urea treatment - is fed to the crossbred cows of the cropfarmer. If in the past, the untreated straw was fed to the local cows of the labourer, or used as bedding, the Simon effect occurs, ultimately resulting in equity problems (Ch. 2.2).

Secondly, a higher total system output is often achieved with less animals of higher individual output when feed quality increases. In systems where farm size does not allow further reduction of farm system herd size, it implies that fewer farmers will produce higher output, 'thanks' to the externalization of weaker farmers: an important social trade off!

Simplification and artefacts

At the risk of introducing biases both in favour and opposed to the use of straw and low individual animal output, the thought experiments were dramatically simplified. These biases and possible artefacts merit some discussion in order to better interpret the results.

The expression of animal output (= maintenance + milk) as multiples of M has a major advantage: it reflects animal system productivity in terms of feed by comparing output over input in terms of feed (not counting the value of products such as dung and saving account

---

[3] The terminology of weak and strong (sub)systems is maintained here, but it needs further definition. Push and pull is difficult to distinguish, i.e. a strong system can either push its excess into, or extract its shortage from a weaker system. However, when extraction is done to meet the requirements of the damning objective, it can be said that resources are extracted from weaker systems.

however, nor investments and costs of labour that are strictly speaking also part of the maintenance requirements). Thus it avoids the impression that a large animal is more 'productive' in terms of feed use than a small animal, simply because its larger frame produces more milk. Generally, the maximum DMI of small and large ruminants, being directly related to animal output in terms of milk, or liveweight gain, is a fixed ratio when expressed in terms of metabolic body weight (Taylor, 1980b). This ratio is reflected in our choice of a maximum production of 3M (= 1M for maintenance + 2M for milk!). A different ratio would not affect the reasoning of this paper. The disadvantage of this simplification lies in the neglect of differences between animal species or breeds, a difference that can be large indeed (Coppock *et al.*, 1986; Hofmann, 1989; NRC, 1991). The issue does not greatly affect the points of this paper but it excludes the possibility of employing variation between animals to increase system output. The point appears in farmers' practice, where feed refused by high milk producers (see Figures 2Af and 2Bf), is used for lower producers, made more attractive where even animals around maintenance still produce dung, a point brought out more explicitly in the modelling of Zemmelink *et al.* (1992). The same principle was shown in the work of one of our students (Kater, 1989), who used more than one type of cow in the thought experiments. She showed clearly that the model maximizes system output by using both low and high producers. It represents a third case, not further discussed here, where not only the feed is desaggregated, as in case II, but where also the type of animal would be desaggregated.

The use of 0.75*M animals, i.e. sub-maintenance levels of production, implies a negative milk production of -1.5 litres (Table 1), valued in our objective function as 0.001 lts to ensure that the model 'recognizes' that cow. Our use of such a value stresses that dung, draught and survival can be an essential form of production, particularly in LEIA. The point shows up particularly in multiperiod planning, where straw can be used as a feed in the lean season (Insiani, 1990), in the modelling by Zemmelink *et al.* (1992), and again by farmers' practice in many if not all LEIA systems. By attaching a value to the use of dung, draught and survival, all these models ensure that lower quality feeds are valued more highly than when only the milk production is considered useful.

One notable problem is caused by uncertainty about the value for F_p, i.e. the relation between increased intake and the level of production. As F_p is allowed to increase, it also increases absolute levels of system output in terms of milk (Figures 2Bp and 2Dp in Figure 4). However, the value or F_p does not affect the principle of the damning objective. This type of issues, and the effect of exclusion of integers, multiperiod planning, changing herd composition and whole farm planning warrants more work, but without much difficulty it should be possible to further illustrate the point of Columella in Roman times, that *'the cow should be allowed to calve every second year ....'*. In other words, there are conditions that high individual milk or calf crops are counterproductive for total system output!

Assumptions that favour or disfavour the use of straws

On the one hand, the exclusion of requirements for milk consumed by calves, growth and pregnancy, favour system output from cows of lower individual output. Also the fact that the cost of labour and animal housing is not counted, favours the use of animals with low individual output in terms of milk. Moreover, the approach to use all feed for animals, ignores the essential use of feed biomass for non-feed purposes, e.g. soil organic matter, thatching or paper manufacture, favouring the use of straw for feed, and thus the use of animals near maintenance.

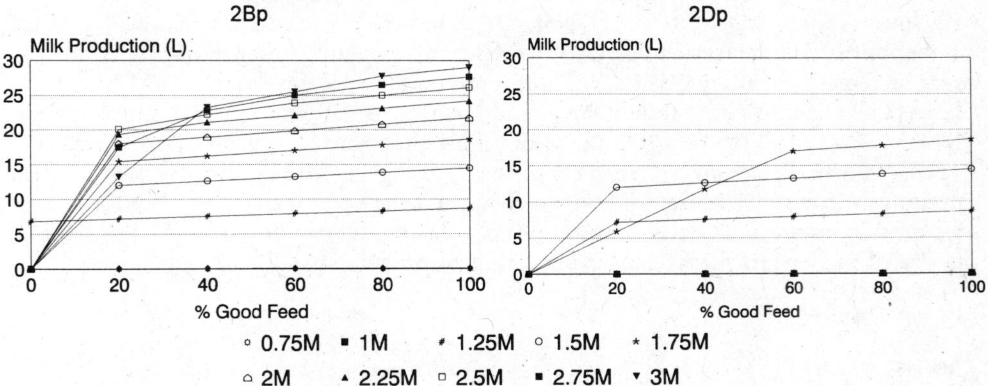

Figure 4:   The effect of F_p on the magnitude of the damning objective (case IIb and IId)

However, the value of straw and animals around maintenance is underestimated by not including seasonal effects. Straw and mature grasses can be essential for herd survival in lean seasons, to allow animal production in flush seasons (Insiani, 1990; Ifar *et al.*, 1995). Furthermore, Ch. 4.2 showed that calculated performance (P_{calc}), overestimated the real performance (P_{real}), thus overestimating the output from the higher producers. Also, the marginal productivity of nutrients is assumed to be equal for the production of milk in low- and high-producers, and maintenance requirements of high- and low-producers are assumed to be equal. These are both doubtful propositions that favour the high producers, though 'sanctioned' by the NRC tables upon which our work is based. Moreover, maintenance requirements include not only the nutrients from feeds for animal metabolism, but also those for housing, veterinary care and market infrastructure (Ch. 6). The possibility of lower maintenance requirements for animals or breeds, at a generally lower level of output is suggested by Frisch and Vercoe (1978), but here ignored, again a bias against the value of straw. The underestimation of the value of straw is aggravated by exclusion of the use of feed refused by high-producers for near-maintenance-producers. Last but not least, the model favours high producers by allowing a higher intake of a mix of good and bad feed than is strictly permitted by the formula for dry matter intake. It assumes that the intake of a mix of good and bad feed is equal to the intake of the good feed alone, thus allowing more high producers and a higher total system output that what should be possible in practice.

Miscellaneous aspects

Repeated runs allow the use of a simple matrix, here using only 5 variables and 6 constraints. Morrison *et al.* (1986) used a similar approach and considered a matrix of 290 variables by 130 constraints still to be manageable. Clearly, with additional variables and constraints, this small matrix will also rapidly expand. The advantage of a simple matrix has to be weighed against the need to adjust coefficients for successive runs. The number of runs can be reduced by deleting ranges that give either a non-feasible solution (e.g. exceeding 1.25M at a feed quality less than TDN55/CP10 in case IA), or by deleting ranges with linear responses (e.g. from TDN55/CP10 and higher for 1.25M in case IA). The number of runs can be further decreased by reducing the number of increments. However, larger increments reduce precision, and introduce the slope of lines such as MM' and AA'. Theoretically, those lines should be vertical since e.g. at one single point between TDN55/CP10 and TDN60/CP12 the solution becomes zero for a 2.25M cow in case IA. Another drawback of large increments is that e.g. Figure 1, they wrongly suggest that solutions for cows of 1.5-2.25M become zero at the same point (i.e. TDN55/CP10). These problems do not invalidate the points in this paper, and they can be overcome by reducing the size of increments for feed quality and animal productivity (Figure 5).

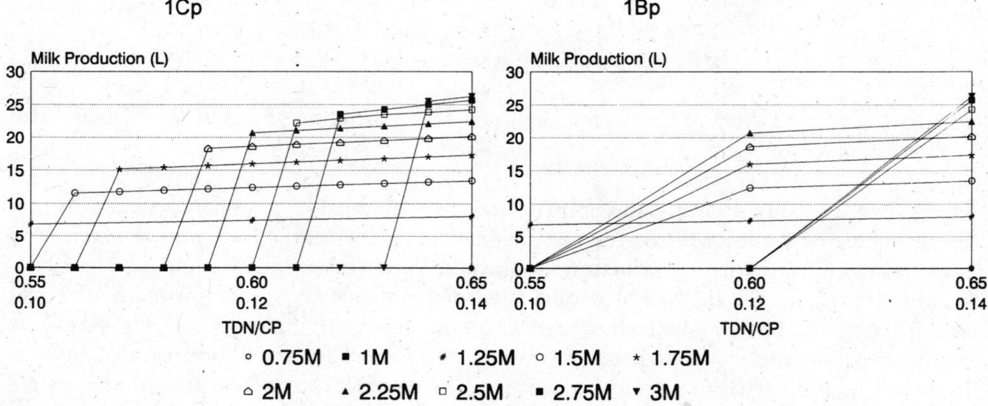

Fig 5: The effect of smaller increments on the shape of the output figures, calculated for ten (case Ic) rather than two (case Ib) increments on the scale from TDN55/CP10 to TDN65/CP14

The use of simple feed balances, as an aggregate summation of nutrient availability and animal requirements, is misleading in a number of ways. First, the setting of excessive production targets, i.e. damning objectives precludes the use of locally available feed resources, and necessitates the use of inputs. Secondly, aggregation of feeds and requirements, excludes the possibility of selective consumption, i.e. instead of trying to feed all feeds, there can be an advantage in deleting part of the feed resource. Also, though not elaborated here, a simple feed balance tends to mask a protein or energy excess in one animal category, with a deficiency in another animal category. The problem of indivisibility of production factors is not discussed, i.e. our results ignore the existence of an optimum farm size at a given point of time, or the fact that a farm system cannot realistically own a part of an animal. The issue can be solved mathematically with integer planning, but in practice, the partial cow is replaced with one or more small animals if no additional feed can be produced or imported. Otherwise, but more seriously in terms of equity (Conway, 1985), the feed for the partial cow will be either discarded, used for other purposes or for the feeding of a partial animal in another, generally stronger, farm system. Total system output thus increases at the cost of a social trade-off, and marginalization of farmers becomes likely. The logic that this frees labour for other functions in society (Boserup 1965) is attractive, but the practice, again in low input conditions, is that many of these farmers are not likely to find employment in situations of structural underemployment or chronic lack of access to resources (Lele and Stone, 1989).

CONCLUSIONS

Changing resource / demand patterns require changes in farming system design. Since livestock in low-input mixed systems are often an essential but secondary subsystem to the crop subsystem, there are limited possibilities to cultivate fodder or to obtain feed from outside. The results of thought experiments, where different types of animals and feeds are combined for maximum system output under these LEIA conditions, indicate that under a given feed supply, the animal output targets have to be adjusted to the feed availability on the farm, in order to realize higher system output.

In all cases, the system output expressed in terms of milk, increases with feed quality, mainly achieved by reducing the number of animals. Unfortunately, this represents a trade-off between the equity of cattle ownership and total system output. Though much affected by the rigidity of the target, an excessive targeting for high subsystem output, such as individual animal yield, negatively affects overall system output, a principle caused by what we propose to call the damning objective. The theoretical possibility of calculating an ideal production target for achieving maximum system output, provides an option to predict whether a farm system should consist of local cows, crossbreds or purebreds, depending on the rigidity of the output from these animals. Desaggregation of feed pools, i.e. the selective consumption allows an increase of the total system output in terms of milk by feeding less than what is really available. As with the improvement of feeds, it allows a combination of more animals whith a higher output per animal.

The models used here are highly simplified, and they express LEIA system strategies. Further work should focus on the effect of integer planning, multiperiod planning, refinement of the software, the relation between genetic production potential and feed intake, the inclusion of herd composition and allowances for pregnancy, milk consumption by calves and the effect of fixed costs in terms of economics. Important effects of subsystem adjustment and resource distribution for maximum output on problems of equity, i.e. access to production and resources, have become apparent and deserve further analysis.

ACKNOWLEDGEMENTS

This paper reflects the work and discussions particularly with senior colleagues A. Zurek, B.R. Patil and B.O. Brouwer, and with students such as Rudolf Pieterson, Loes Kater and Marinus Bos. The valuable suggestions of B.J. Tolkamp, G. Zemmelink and N.W.M. Ogink are very much appreciated.

REFERENCES

Alexander, G.I., Reason, G.K. and Clark, C.H., 1984[a]. The development of the Australian Fiesian Sahiwal. A tick-resistant dairy breed. World Anim. Rev., 51: 27-34.

Alexander, G.I., Reason, G.K., Gale, G.M.R. and Clark, C.H., 1984[b]. The performance of Australian Friesian Sahiwal Cattle. Darwin, northern Australia. World Anim. Rev., 52: 13-16.

Allden, W.G., 1970. The effect of nutritional deprivation on the subsequent productivity of sheep and cattle. Nutr. Abstr. Rev., 40: 1167-1184.

Breman, H. and De Wit, C.T., 1983. Rangeland Productivity and exploitation in the Sahel. Science 221 (4618), p.1341-1347.

Boserup, E. 1965. The conditions of agricultural growth. The economics of agrarian change under population pressure. George Allen & Unwin Ltd., London, UK. 124 pp.

Conway, G.R., 1985. Agroecosystem analysis. Agric. Admin., 20: 31-55.

Conway, G.R., and Barbier, E.B., 1990. After the Green Revolution, Sustainable agriculture for Development, Earthscan Publications Ltd. London, 205 pp.

Coppock, D.L., Ellis, J.E., and Swift, D.M., 1986. Livestock feeding ecology and resource utilization in a nomadic pastoral ecosystem. J. of Appl. Ecol., 23: 573-583.

Crotty, R. 1980. Cattle, Economics and Development. Commonwealth Agricultural Bureaux, Farnham Royal, Slough Sl2 3BN, UK. 253 pp.

Daly, H.E., and Cobb, J.B., 1990. For the common good. Redirectiong the Economy towards Community, the Environment and a Sustainable Future. Green Print. 482 pp.

De Wit, C.T., 1992. Resource use efficiency in agriculture. Agric. Syst., 40: 125-151.

De Wit, J., Dhaka, J.P. and Subba Rao, A., 1993. Relevance of breeding and management for more or better straw in different farming systems. p.404-414. In: Kiran Singh and Schiere, J.B., (eds.), 1993. Feeding of ruminants on fibrous crop residues. Aspects of treatment, feeding, nutrient evaluation, research and extension. Proc. of a workshop, 4-8 february 1991, NDRI-Karnal, ICAR, New Delhi, India. 486 pp.

DGIS, 1987. Assisting Livestock Development, Experience of development Cooperation with reference to livestock in the period 1978 - 1984. Evaluation Report Netherlands Development Cooperation. 80 pp.

Durning, A.B., and Brough, H.B., 1991. Taking stock: Animal Farming and the Environment. World Watch Paper 103.

Forbes, J.M., 1986. The voluntary food intake of farm animals. Butterworth, London, UK. 206 pp.

Fraser, A., 1949. Sheep husbandry. London, UK. 298 pp.

Frisch, J.E. and Vercoe, J.E., 1978. Utilizing breed differences in growth of cattle in the tropics. World Anim. Rev., 25: 8-12.

Gibson, T., 1987. A ley farming system using dairy cattle in the infertile uplands, Northeast Thailand. World Animal Review, 61: 36-43.

Hayami, Y. and Ruttan, V.W., 1985. Agricultural Development, an International Perspective. The John Hopkins University Press, Baltimore and London. 506 pp.

Hayman, R.H., 1974. The development of the Australian Milking Zebu. World Anim. Rev., 11: 31-35.

Hofmann, R.R., 1989. Evolutionary steps of ecophysiological adaptation and diversification of ruminants: a comparative view of their digestive system, Oecologia, 78: 443-457.

Ifar, S., Solichin, A.W., Udo, H.M.J. and Zemmelink, G., 1995. Subsistence farmers' access to cattle via sharing in the limestone area in the Southern part of East Java, Indonesia. In Print

Insiani, Y., 1990. Straw / grain ratio in rice straw as related to animal production. MSc-thesis, Dept. Trop. Animal Production, Agricultural Univ. Wageningen, The Netherlands.

Johnston, A.E., 1972. The effects of ley and arable cropping systems on the amounts of soil organic matter in the Rothamsted and Woburn ley-arable experiments. p.131-159. In: Rep. Rothamsted Exp. Stn. for 1972, part 2.

Joshi, A.L., Doyle, P.T. and Oosting, S.J., (eds.), 1994. Variation in the quantity and quality of fibrous crop residues. Proceedings of a national seminar held at the Baif Development Research Foundation, Pune (Maharashtra-India), February 8-9, 1994. BAIF, Pune, and ICAR, Krishi Bhawan, New Delhi, India. 174 pp.

Kaasschieter, G.A., De Jong, R., Schiere, J.B. and Zwart, D., 1992[a]. Towards a sustainable livestock production in developing countries and the importance of animal health strategy therein. The Veterinary Quarterly, 14 (2): p.66-75.

Kaasschieter, G.A., Schiere, J.B. and De Wit, J., 1992[b]. Crop - livestock integration a model approach with special reference to different parameters for sustainability. p.341-346. In: Gibon, A., and Matheron, G., 1992, (eds.), 1992. Agriculture, Agrimed research programme; Global appraisal of livestock farming systems and study of their organizational level: concepts, methodology and results. Proc. of a symposium organized by INRA-SAD and CIRAD-IEMVT, 7 July 1990 at Toulouse, France. European Communities, Fr-En.

Kater, L., 1989. Een alternatief model: een kwantitatieve en kwalitatieve analyse van een laagland bos- en savanne landbouw systeem. MSc-thesis, Agricultural University Wageningen, Dept. Trop. Animal Production, Wageningen, The Netherlands.

Kelley, J., 1992. Upgrading of waste cereal straws. Outlook on Agriculture, 21 (2): 105-108.

Ketelaars, J.J.M.H. and Tolkamp, B.J., 1992. Toward a new theory of feed intake regulation in ruminants. 1. Causes of differences in voluntary feed intake: critique of current views. Livestock Production Science, 30: 269-296.

Kingwell, R.S. and Pannell, D.J., (eds.) 1987. MIDAS - a bioeconomic model of a dryland farm system. Simulation Monographs, Pudoc, Wageningen, The Netherlands.

Kiran Singh and Schiere, J.B. (eds.) 1993. Feeding of ruminants on fibrous crop residues. Aspects of treatment, feeding, nutrient evaluation, research and extension. Proceedings of a workshop held from 4-8 February 1991 at NDRI-Karnal. ICAR (Animal Sciences), New Delhi, India. 486 pp.

Lele, U. and Stone, S.W., 1989. Population pressure, the environment and agricultural intensification, variations on the Boserup hypothesis, MADIA discussion paper 4. 79 pp.

McFarland, A.M.S., Kothmann, M.M., and Blackburn, H.D., 1992. Calibrating a diet selection model for sheep grazing rangelands. Agric. Syst., 39: 361-386.

Morrison, D.A., Kingwell, R.S., Pannell, D.J. and Ewing, M.A., 1986. A mathematical programming model of a crop-livestock farm system. Agric. Syst., 20: 243-268.

Nou, J., 1967. Studies in the Development of Agricultural Economics in Europe. Almquist & Wiksells Boktryckeri AB, Uppsala.

NRC, 1988. Nutrient Requirements of Dairy Cattle, National Research Council, 6th. revised edition. Washington DC, USA.

NRC, 1991. Microlivestock, little-known small animals with a promising economic future. National Academy Press, Washington, D.C.

Odum, H.T., 1983. Systems ecology, an introduction. Whiley Interscience, New York. 644 pp.

Ogink, N.W.M., 1993. Genetic size and growth in goats. Ph.D. Thesis, Agricultural University Wageningen, The Netherlands.

Robertson, D.E., 1983. Defining selection objectives. p.485-492. In: Haresign, W. (ed.), 1983. Sheep Production. Butterworths, London. 576 pp.

Romero C. and Rehman, T., 1989. Multiple criteria analysis for agricultural decisions. Developments in agricultural economics 5, Elsevier, Amsterdam, The Netherlands.

Renkema, J.A., 1972. De opbouw van lineaire programmeringsmodellen ten behoeve van de agrarische bedrijfsplanning. Ph.D. Thesis. publikatie nr.4., afdeling voor agrarische bedrijfseconomie aan de Landbouwhogeschool, Wageningen, The Netherlands.

Schiere, J.B., 1992. Aspects of modelling for sustainable crop-livestock integration. p.433-443. In: Ibrahim, M.N.M., De Jong, R., Van Bruchem, J. and Purnomo, H. (eds.), 1992. Livestock and Feed development in the tropics. Proc. of the Int. Seminar held from 21-25 October, 1991 at Brawijaya University, Malang, Indonesia. 546 pp.

Simmonds, N.W., 1986. A short review of farming systems research in the tropics. Exp. Agric., 22: 1-13.

Sol, J., Renkema, J.A., and Brand, A., 1984. A three year herd health and management program on thirty Dutch dairy farms. IV. Special aspects. The Veterinary Quarterly, 6 (3): 163-169.

Spedding, C.R.W., 1987. An introduction to Agriculture Systems, 2nd Edition, Elsevier, London.

Staniforth, A.R., 1982. Straw for fuel, feed and fertilizer. Farming Press Ltd., Wharfedale Road, Ipswich, Suffolk, UK. 153 pp.

Sundstøl, F., and Owen, E., (eds.), 1984. Straw and other fibrous by-products as feed. Developments in Animal and Veterinary Sciences, 14. Elsevier, 604 pp.

Taylor, St.C.S., 1980. Liveweight growth from embryo to adult in domesticated mammals. Anim. Prod., 31: 233-235.

Udo, H.M.J. and Brouwer, B.O., 1993. A computerised method for systematically analysing the livestock component of farming systems. Computer and Electronics in Agriculture, 9: 335-356.

Van Niejenhuis, J.H. and Renkema, J.A., 1989. De opbouw van modellen ten behoeve van de mathematische programmering van agrarische bedrijven. (the construction of models for the mathematical programming of agrarian enterprises), Wageningse Economische Studies: 15, Agricultural University Wageningen. Pudoc, PO-box 4, 6700 AA Wageningen, The Netherlands. 171 pp.

Wahed, R.A., Owen, E., Naare, M. and Hosking, B.J., 1990. Feeding straw to small ruminants. Effect of amount offered on intake and selection of barley straw by goats and sheep. Anim. Prod., 51: 283-289.

White K.D., 1970. Roman Farming. Thames & Hudson, London.

Woodward, L., and Foster, L., 1988. The use of herbal leys in modern British organic farming systems. p.421-431. In: Allen, P. and Van Dusen, D. (eds.), 1988. Global perspectives on agroecology and sustainable agricultural systems. Proc. of the sixth Int. Scientific Conference of the International Federation of Organic Agriculture Movements. Vol.1.

Zemmelink, G., 1980. Effect of selective consumption on voluntary intake and digestibility of tropical forages. Ph.D. Thesis. Agric. Res. Rep. 896, PUDOC, Wageningen, The Netherlands, 100 pp.

Zemmelink G., 1986. Measuring intake of tropical forages. p.17-21. In: Balch, C.C., and Van Es, A.J.H. (eds.) 1986. Recent advances in feed evaluation and rationing for dairy cattle in extensive and intensive production systems. International Dairy Federation, Brussels, Bulletin No.196

Zemmelink, G., Brouwer, B. and Ifar, S., 1992. Feed utilization and the role of ruminants in farming systems. p.444-451. In: Ibrahim, M.N.M., De Jong, R., Van Bruchem, J. and Purnomo, H. (eds.) 1992. Livestock and Feed development in the tropics. Proc. of the Int. Seminar held from 21-25 October, 1991 at Brawijaya University, Malang, Indonesia. 546 pp.

Section 6

CATTLE, STRAW AND SYSTEM CONTROL

DISCUSSION

'O speculators about perpetual motion, how many vain chimeras have you created in the like quest? Go and take your place with the seekers after gold.'
Leonardo da Vinci, quoted by: Partington, J.R., 1961 A History of Chemistry, volume two. Macmillan, St. Martin's Press, London. 795 pp.

'One of the main subjects in present-day physics is the problem of elementary particles. However, we know that elementary particles are far from elementary. New layers of structure are disclosed at higher and higher energies. But what, after all, is an elementary particle? Is the planet earth an elementary particle? Certainly not, because part of this energy is in its interaction with the sun, the moon, and other planets. The concept of elementary particles requires an "autonomy" that is very difficult to describe in terms of the usual concepts. Take the case of electrons and photons. We are faced with a dilemma: either there are no well-defined particles (because the energy is partly between the electrons and protons), or there are noninteracting particles if we can eliminate the interaction. Even if we knew how to do that, it seems too radical a procedure. Electrons absorb photons or emit photons. A way out may be to go to the physics of processes.'
Ilya Prigogine and Isabelle Stengers, 1985, in "Order out of Chaos; Man's New Dialogue with Nature". Flamingo, London, page 287-288.

Chapter 6

CATTLE, STRAW AND SYSTEM CONTROL:
A DISCUSSION

This thesis addresses two related questions. The first one concerns the *'suitability of straw feeding methods in mixed crop-livestock systems'*, and it focuses on the use of urea ammonia treatment of straw, with or without concentrate supplementation. The second question concerns *'the role of straw in the drive and shape of systems'*, in an attempt to discern a logic between system development and usefulness of different straw feeding methods. This discussion reviews the conclusions of previous chapters and it verifies the results with concepts from other branches of science. The chapter consists of five parts and an epilogue. The first part reviews concepts and methodologies from Farming Systems Research (FSR) as they were applied in the study of the usefulness of straw feeding methods. The second part summarizes the results of animal feeding trials, i.e. component research, that together with a set of 'thought experiments' determine the suitability of feeding treated straw as a medium quality forage. The third part explains laws that govern system morphogenesis. It uses concepts from thermodynamics and information theory, branches of science that study the role of energy and information in system control. It also elaborates the issue of criteria for system success that, together with the resource availability, form the resource / demand patterns, i.e. the system boundary conditions, the major determinants of the system morphogenesis. The fourth part reviews some general aspects of technology development and subsystem adjustment for maximum system output, here called the issue of the communal ideotype. The final part ties it all up in a discussion on the emerging logic between the usefulness of straw feeding systems and the mode of farming. While doing so, this last chapter also proposes tentative explanations for issues that were mentioned, but not elaborated in the previous chapters, e.g., punctuated development, trade offs, boundary conditions, damning objectives and Simon effects, as well as problems of equity. All these points started to make more sense as the work progressed. They, therefore, deserve special mention in this discussion, even though their full explanation requires cooperation between disciplines beyond animal nutrition alone, an approach that is a typical precondition for successful farming systems research.

FARMING SYSTEMS RESEARCH AND STRAW FEEDING

The first chapter is a review on methodology and backgrounds of FSR in a broad sense, i.e. FSR *sensu latu* (Simmonds, 1986). It provided analytical tools that were useful to the work on both questions of this thesis, and it raised interesting issues for further research (Ch. 2.1). Firstly, the review explained the use of thought experiments and agroeco-zoning, concepts that have facilitated the understanding of system behaviour in this thesis. By abstraction and simplification they permitted the study of systems and questions that are beyond experimentation (Ch. 2.2, 4.1 and 5.2). Secondly, the review discussed issues such as the definitions of systems and farming, it identified similarities between cropping and livestock system research (CSR and LSR), and it justified the need to pursue FSR in spite of the large amount of past work. Thirdly, and more important for this discussion, the review identified fundamental issues in FSR, e.g., definitions of system boundaries, and the similarity in behaviour between systems at all levels of the hierarchy.

Agroeco-zoning, resource / demand patterns and system boundaries.

As the work on the first question, i.e. the usefulness of straw feeding methods progressed, the need for a logical system classification assumed greater importance. This led almost naturally to the second question i.e. about the role of straw in the drive and shape of systems. The attempt at classification required a form of agroeco-zoning, an activity that defines farming systems according to agroecological criteria. Though often not explicitly mentioned, not only agroecological, but also sociological, cultural and economic criteria need to be used to form a realistic classification. The resulting multitude of criteria, however, forces the researcher to simplify, abstract and summarize these criteria, according to the scope and objective of the work. According to Ch. 2.2 and 2.3, the logic of system behaviour can be simplified to be determined by the access to the classical production factors land, labour and capital, in relation to the effective demand for farm output. In fact, the work on the effect of these so-called resource / demand patterns on the development of farming systems, took away the time and energy to study components and relations within subsystems, an approach that is more prevalent in the work by Shaner *et al.* (1982) and Odum (1983). Deliberately, this study then focused on processes rather than on the details of individual farm systems.

Rather independently of each other, both the work on the classification of farming systems in Ch. 2.2 and the discussion on applicability of western feeding standards in tropical systems (Ch. 5.1) concluded that systems can be distinguished in what can be called 'open' and 'closed' systems (box 1). On that basis, it can be hypothesized that the 'degree of openness', e.g., the access to the market and the possibility to exploit more land, or other resources, reflects the resource / demand pattern, which in turn determines system morphogenesis. The degree of openness thus reflects what can also be called the boundary conditions of a system. This terminology and approach occupies a central place in the study of 'chaotic behaviour' of non-linear systems. It is a relatively new branch of science that studies, *inter alia*, the rules of what can be called system morphogenesis, i.e. the development and shaping of a system over time (Prigogine and Stengers, 1985; Gleick, 1987; Lewin, 1993).

The second question of this thesis, i.e. about the drive and explanation of system morphogenesis came almost naturally with the first question. The study of system morphogenesis is not new, and it will be the main topic in the later parts of this discussion. Interestingly, it has been important in studies on evolution of biological systems for quite some time (Darwin, 1859; Eiseley, 1957; Dawkins, 1991), but it now appears to be equally relevant, for the evolution of farm systems, or possibly, many other systems. This notion of an evolution is unavoidable indeed, after identifying and arranging a large number of farm systems on scales of resource / demand patterns that can be supposed to determine system behaviour (Ch. 2.2). As access to resources change, relative to the demand, the systems develop 'technologies' for survival, whether by chance and by learning. With the benefit of hindsight, it has become clear that this classification was bound to also serve in the discussion of the thermodynamic aspects of system development, as explained later in this chapter. Development in this sense is not seen as a one way direction from bad to good, it is merely a response to changing resource / demand patterns where *necessity is the mother of invention*. It can be seen as the result of a process of the survival of the fittest, a form of system evolution that applies 'induced innovations' (Hayami and Ruttan, 1985), with a Lamarckian notion that 'induced innovations' become 'acquired characters' (see for example Maynard Smith, 1982, 1989; Dawkins, 1982, who also quotes Cavalli-Sforza and Feldman, 1981).

<table>
<tr><td>

Box 1: SOME NOTES ON THE TERMINOLOGY 'OPEN' AND 'CLOSED' SYSTEMS

The distinction between open and closed systems might require another terminology, to be thermodynamically correct. Indeed, no agricultural system is closed, since solar energy is entering per definition. Moreover, and even if a system does not actively import additional energy, it may be used as a dumping ground for waste from other systems. Without energy, a system attains *'equilibrium'*, a state in which no agriculture is possible. If little energy is entering, i.e. in systems that have limited access to resources, it may be better to speak of systems *'near equilibrium'*. An extreme case of such systems are the arctic lakes of Canada described by Johnson (1981), but LEIA might be hypothesized to fit the same pattern. Systems with high energy inputs, e.g., HEIA, are probably better called *'distant from equilibrium'*. These concepts are well explained by Prigogine and Stengers (1985), and though these points remain to be proven, they might present an exciting field of further study in system morphogenesis.

</td></tr>
</table>

Thought experiments and punctuated development

Thought experiments, a form of modelling, help to understand, explain and predict system behaviour on ranges of resource / demand patterns and levels of complexity that defy experimentation. They were used successfully in Ch. 4.1 and 5.2, even though they may indicate trends rather than absolute values. The impression of a predetermined and mechanistic view of development, which allows prediction of 'system trajectories', is at least partly defused by the same 'Chaos theory' that has been mentioned earlier (Prigogine and Stengers, 1985; Gleick, 1987). In those concepts, the principle of the extreme dependency on initial conditions explains how a minute change in the initial state of a system can unpredictably affect the behaviour of non-linear systems. This so-called butterfly effect takes place particularly when negative feedback mechanisms are absent. The details go beyond this thesis, but the topic provides a challenging area for research into the morphogenesis of farm systems, for example, whether gradual or punctuated development is likely and / or necessary. The issue might be particularly relevant for the study of (farm)systems in high external input agriculture (HEIA), a mode of agriculture that tends to cancel or at least to delay feedback mechanisms, as they adjust the resource use to the objective (Odum, 1971; Ch. 2.2 and 5.1). The work by Meadows *et al.* (1972) gives typical examples of drastic system change when feedbacks are delayed.

Paradoxically, the principles of this 'Chaos theory' reduce the reliability of prediction due to punctuated, chaotic behaviour, while also providing a theoretical background as to why different initial conditions can lead to similar states, so-called attractors. The possibility of recurring shapes, or stereotypical farm systems in farm system development is clear from the examples in the farm(ing) system classification matrix of Ch. 2.2. This matrix is a form of agroeco-zoning that also provides a framework for the answer to the first question of this thesis. Together with the use of thought experiments, it helps in the identification of stereotypical farming systems where different straw feeding systems might fit. Cynically, this approach can be called a form of Farming Systems Research / Extension (FSR/E) in reverse, where:

the technology is known but where the problem remains to be identified.

COMPONENT RESEARCH:
STRAW TREATMENT AND SUPPLEMENTATION

Thought experiments need data and models, besides independent data sets or analogies and laws from other sciences for their verification. The thought experiments of Ch. 4.1 determine the economics of feeding urea treated straw with or without supplement by using a simple LP-model based on data from animal trials reported in the literature. At the same time, a series of on-station trials were started to create an additional and independent data set on the nutritional parameters and animal responses (Ch. 3.1-3.4). These trials showed that, under the conditions of these experiments:
- urea treatment consistently increases straw dry matter intake and digestibility in large ruminants, if straw is a large part of the ration,
- substitutional supplementation results in rather linear responses when seen over a large range of supplementation,
- the effect of catalytic supplementation based on associative effects is smaller than of urea treatment, it is unpredictable and probably only relevant around maintenance levels of production (See Ch. 2.3 for technical terms and Ch. 4.2 for nutritional backgrounds).

Thought experiments on suitability
of treated straw feeding

The thought experiments of Ch. 4.1 further identified the type of farm systems that might profitably adopt the use of urea treated straw. They help to answer the first question of the thesis by concluding that urea treatment of straw is a 'niche solution', i.e., its application is system specific. Compared, and combined with supplementation, urea treatment is useful where
- plenty of straw is available relative to other feeds, i.e. where straw is cheap compared to green feeds or concentrate supplements,
- the level of animal production ranges between low and medium, or when expressed in

multiples of maintenance between approximately 1.5 to 2.5*M, with one time maintenance for animal survival included, depending on the price ratios of straw, supplement, milk and meat,
- the cost of feeding treated straw can be recovered from the sale of products such as milk or meat,
- urea, water and covering materials are cheap, i.e. easily available.

All such conditions for success of straw treatment make sense for those with some understanding of animal feeding systems. The need for access to the market, for the sale of products and the purchase of inputs, may seem too obvious for words. However, this very need for exchange with other systems is often overlooked, and it highlights again the importance of boundary conditions.

Testing of results

The testing of the results about the usefulness of straw feeding methods was done from a nutritional angle in Ch. 4.2, from a socio-economic angle in Ch. 5.1, and against field observations while visiting a variety of farming systems. The conclusions are repeatedly confirmed in reports such as by Westgaard and Sundstøl (1986) and in Kiran Singh and Schiere (1993). Contacts with farmers reconfirm the conclusions over and over again, and some examples may serve to make this point:
- *a Sri Lankan farm woman was of the opinion that straw treatment saved time to carry grass, implying that treated straw was cheap in comparison with grass (pers. obs.),*
- *an Indian farmer stopped feeding treated straw because the supplement had become cheaper (A.L. Joshi, pers.comm., 1993),*
- *Scottish farmers do treat straw when the grains are expensive and not when the grains are cheap (pers. obs.).*

The nutritional testing of the results in Ch. 4.2 show that prediction of animal performance with ration formulation helps to correctly rank the response, but it tends to overestimate actual response. Testing of the results from a socio-economic angle in Ch. 5.1 argues that - extremely speaking - in high external input agriculture (HEIA) the feed tends to be adjusted to the objective of animal production. In low external input agriculture (LEIA) the animal output tends to adjust to the feed supply. New Conservation Agriculture (NCA) occupies a position between these two extremes, and it offers scope to combine the concepts and practices of both HEIA and LEIA. The thought experiments can be used for ration formulation in both HEIA and LEIA. They can calculate the cheapest ration for a range of production levels, and subsequently determine the most optimal production level for a given resource / demand pattern. Indeed, depending on the feed supply and fixed costs, the optimum level of animal output can fluctuate. In other words, the system shape can change, determined as it is by boundary conditions, and accordingly, the criteria for system success will change. This principle first became apparent in Ch. 2.2 and it was explored in Ch. 5.2 with additional thought experiments, showing once again how system morphogenesis is determined by resource / demand patterns. This point is a convenient stepping stone to proceed to the next section, and to the discussion of the second question in this thesis.

SYSTEM CONTROL AND THE LINK WITH STRAW AND CATTLE

The research on the second question of this thesis concerns *'the role of straw in the drive and shape of systems'*. This work is a form of FSR *sensu strictu*, an academic form of FSR, that in this thesis became increasingly focused on the distinction between open and closed systems in relation to effective demand. That distinction has provided at least some clues to system behaviour, and a brief explanation of the principles of system control is required here, metaphorically starting in 'paradise'. When 'Adam and Eve', or their colleagues in other parts of the world, wanted - or needed - more than what was allotted to them, they had to start working: in sweat and tears they had to start tilling the soil. Probably, they first went hunting and gathering, but when their effective demand continued to increase they had to actually begin to manage crops and livestock, the start of expansion agriculture: *necessity is the mother of invention*. Ultimately, in HEIA, their descendants run around to organise their effective demand by spending increasing levels of external energy without much attention to the proper use of information (Odum, 1971; Crosby, 1986; Rifkin, 1989; Simmons, 1989; Ponting, 1991). In the concept of this thesis, NCA maximizes the use of both energy and information. Indeed, the discussion on development and progress in Ch. 2.2 has shown that so-called primitive farming systems are capable of providing more and better food to a relatively small population with low per capita effective demand, at less effort than modern systems (Wilkinson, 1973; Cox and Atkins, 1974; Ponting, 1991). This sequence of events reflects a logic in system behaviour that appears to be explained at least partly with concepts from thermodynamics and information theory.

Thermodynamics, information and entropy

Thermodynamics is the science that studies the transfer of energy into movement and organization. The first law teaches that energy cannot be created nor lost, it can only change form. The second law teaches that all systems, if left on their own, tend towards a state of maximum entropy, also called chaos, or lack of order. Leaving a system on its own is a negative way to define system control. In highly simplified terms, the second law implies that energy tends to flow from high to low concentrations. This tendency towards a state of increased entropy is likely to explain much, if not all, of the drive of systems (Odum, 1971; Prigogine and Stengers, 1985; Lyklema, 1991). As long as the perpetual motion machine is not invented, this law will remain valid (Figure 1), and the answer to the first part of the second question of this thesis, about the drive of systems, needs to be sought here. In this context it is tempting, if not compelling, to imagine that the low energy flows (fluxes) in straws permit and cause a lower and/or slower drive of systems than feeds or energy sources with higher fluxes!

Information theory studies the use of information in the organization of systems. Interestingly, and not accidentally, the statistical formula for entropy and the formula for the extent of information are identical (Bok, 1964; Tribus and McIrvine, 1971; Chancellor, 1981; Kramer and De Smit, 1987). It is not so difficult indeed to see the relation between entropy and information since order itself can be considered to be a form of information. Total lack of order, i.e. the state of maximum entropy, carries least information.

Figure 1   A few examples of attempts to design a perpetual motion machine
(Source: Dieterich, 1986).

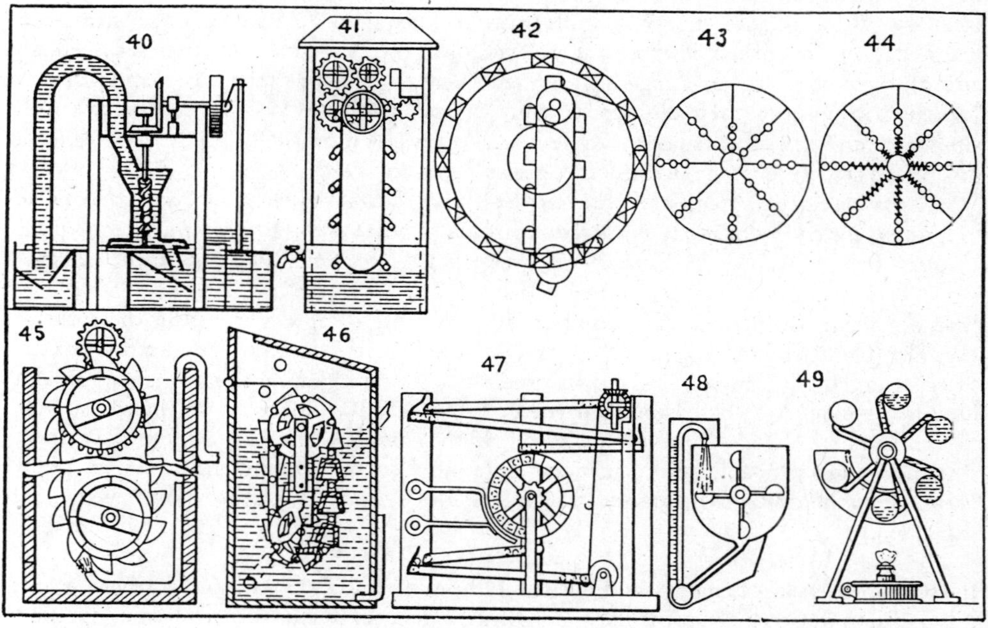

Note: numerous attempts at the design of a perpetual motion machine were - and are still - made, but they
fail due to the rules that are described in the second law of thermodynamics. It is also remarkable
to see that some designs hope to achieve perpetual motion by adding rather than by reducing
complexity.

Entropy and information may seem complicated concepts, but they can be quantified
nevertheless (Tribus and McIrvine, 1962; Chancellor, 1982; Kramer and De Smit, 1987). For
the scope of this discussion it suffices to say that entropy production can be estimated,
simply speaking, both by measuring energy consumption of the system, and by estimating
the degree of system improbability. In principle, the more unlikely or improbable the state
of a system is, the more information it carries, and the more energy is required to achieve
ánd maintain its improbable state. In other words, the more ordered i.e. the more
improbable the shape of a system is, the more energy it requires for its maintenance.
Importantly, a high energy use only implies a high (local!) order, and it needs to be
accompanied by information, achieved at the cost of energy. For example, enzymes as a
form of information reduce the amount of energy required for a reaction. It should be
remembered throughout, however, that according to the second law, a local increase of
order in a sub-system always causes a net decrease of order (increase of entropy) in the
overall system. This is a trade-off with a whole string of possibly important implications
as will be seen later.

System complexity and the cost of maintenance

In simple terms, the more improbable the state of a system is, i.e., the more remote from lack of order, the more energy and information is required for its creation, control and maintenance. This principle applies at all levels of system hierarchy, whether for a cell, an animal, a plant, a farm or a region. It is useful here to remember the observation in Ch. 2.1, that, in spite of their differences, animate and inanimate systems can behave in a similar manner. In that manner, a system in the sense of a unit (Ch. 2.1), can be called a dissipative structure, to be defined as (see also Johnson, 1981):

a system that takes up energy in various forms, that utilizes the energy to perform work, to conserve the energy in newly made structures, and / or releases the remaining energy in the form of heat. By doing so, it transforms energy, and at the same time it dissipates energy.

Live organisms maintain and (re)produce by managing to capture *energy* in the form of sunlight (plants) or food (animals), a process based on *information* supplied by DNA[1]. Equally so, cows, farm systems and also society, exist by ingesting energy in the form of feed or (fossil) energy, based and managed by information from the cow's, the farmer's or the parliament's DNA, libraries and so on. More complex, also called more developed, systems are less probable and require therefore more energy and information. This point is probably the start of the answer on the shape of systems, but other issues need to be explained first.

In spite of the risks of analogy and over-simplification, but aiming to gain further insight, it is useful to illustrate the principle of complexity and energy requirement here at different levels of system hierarchy, while knowing that this is the point where other disciplines should join in. The following cases should serve to make an important point about the shape of systems:
- within the algae, i.e. within one group of plants, the - simple or primitive - prokaryotic blue algae require less energy for maintenance than more complex -or developed - eukaryotic, green algae. This is expressed as the amount of respiration required by species A and B at a light intensity (= energy flux) of zero (Figure 2).
- at a higher level of plants than the algae, the same point seems to be apparent between C3 and C4 grasses (Figure 3a).
- within types of animals in practical livestock production, there is some evidence that so-called tropical cattle (Bos indicus) have 5-10% lower maintenance requirements than the Holstein Friesians (Frisch and Vercoe, 1978). Their point is reflected and discussed in Ch. 5.2, here illustrated in Figure 3b. It is an interesting area for further reflection : the damning effect becomes more pronounced when output targets are more rigid, and when maintenance requirements of low producers are lower than of high producers.
- on a still wider range of organisms, i.e., from protozoa to homoiotherms, the unicellulars require less energy for maintenance (0.018 watts.$W^{-0.736}$) than the progressively more complicated poikilotherms (0.14 watts.$W^{-0.738}$) and homoiotherms (4.1 watts.$W^{-0.739}$) (Reiss, 1989, quoted by Ogink, 1993).

---

[1] the question about the first "creation" of order is around the corner, but beyond the scope of this thesis. For interesting reading, see Dawkins (1991) and Prigogine and Stengers (1985).

Figure 2   The performance of blue and green algae at different energy fluxes.

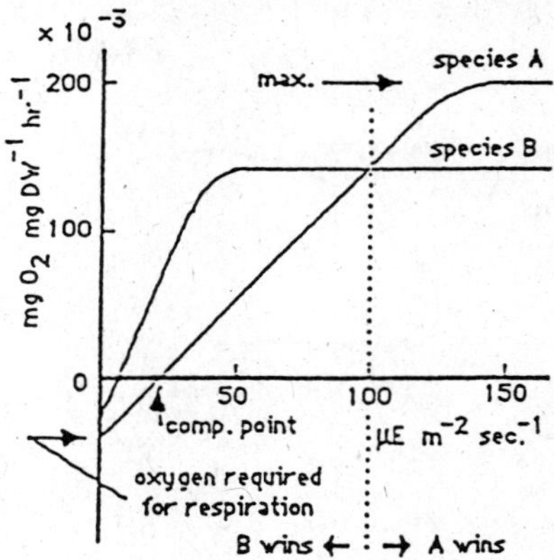

Note: At light intensities > 100 μE m^{-2} s^{-1}, the photosynthetic rate of species A (green algae) is highest, species A will outcompete species B. At light intensities < 100 μE m^{-2}s^{-1}, the reverse will happen, and species B (blue algae) outcompetes species A. The compenzation point and the amount of respiration i.e. energy requirement for maintenance of species B is less than that of species A (Source: Elenbaas, 1994). See Zevenboom, (1986) and Turpin, (1988) for similar but more elaborate comparisons.

Admittedly, some imagination is required, but the evidence is intriguing, when one sees that at an even higher level of system aggregation the same point appears to be valid for agriculture, i.e. for - the shape of - farm systems:

to obtain plants and/or livestock at densities different from those which are typical of the wild ecosystems, humans alter the natural pattern of biota distribution. This 'costs' human society, and can be measured in terms of human labour, fossil energy and technological capital. In principle, the greater the change generated in the natural system to increase the yield of crops and livestock, the greater the flow of power that must be applied by humans. (Giampietro et al. (1992), who quote E.P. Odum (1971) and Stanhill (1984))

The principle of the combination of low maintenance requirements and low outputs is further quantified in a series of energy analysis of farming systems, for example by (Odum, 1971; Leach, 1976; Pimentel and Pimentel, 1979; Stout, 1990; Bayliss Smith, 1991; Netting, 1993; Kessels *et al.*, 1994). Both Figure 4 and Table 1 indicate - though anecdotally - that, as food production and processing becomes more unlikely, its energy efficiency tends to decrease, coinciding with an increased dependence on external fuel sources. The energy ratios need to be interpreted with care, partly because of the difficulty of system definition and boundaries in terms of energy use (Jones, 1989). Nevertheless, and in spite of other evidence by De Wit *et al.* (1987), their point agrees too much with the 'logic' of the farm systems classification in Ch. 2.2, to be ignored. In that classification, which appears to set the path of farm system evaluation, the farm systems move along the vertical axis. As they do so, they tend to also assume increasingly complex shapes, also becoming increasingly dependent on external inputs. Development can be caused by a need, at the expense of (hidden) resources, sometimes leading to an illusion of progress, probably the largest Archimedes principle in both expansion agriculture and HEIA systems (Ch. 1 and 2.2).

Figure 3a   Maximum recorded yields (tonnes dry matter ha⁻¹yr⁻¹) in different C3 and C4 crops at different energy fluxes (source: Cooper, J.P, 1975 (ed.), quoted by Bayliss-Smith, 1991)

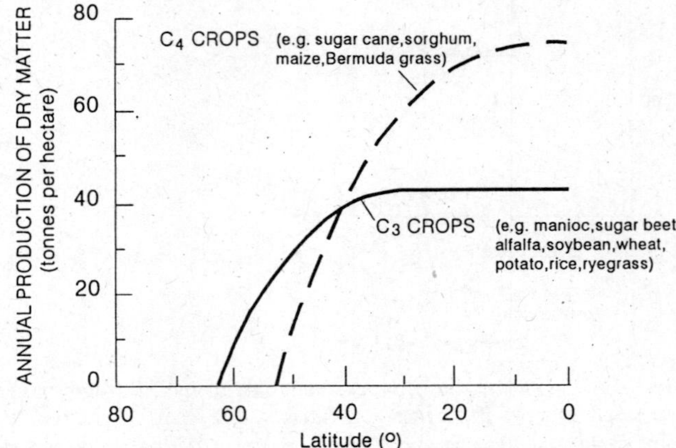

Note: The flux is decreasing at the X-axis since from left to right the latitude, i.e. the distance to the equator increases.

Figure 3b   The performance of a system expressed as litres milk at different energy fluxes, based on subsystems that are low and medium producing animals, where the population of low and high producing cows might be compared with the population of algae in Figure 2 (adapted from Figure 5 in Ch. 5.2).

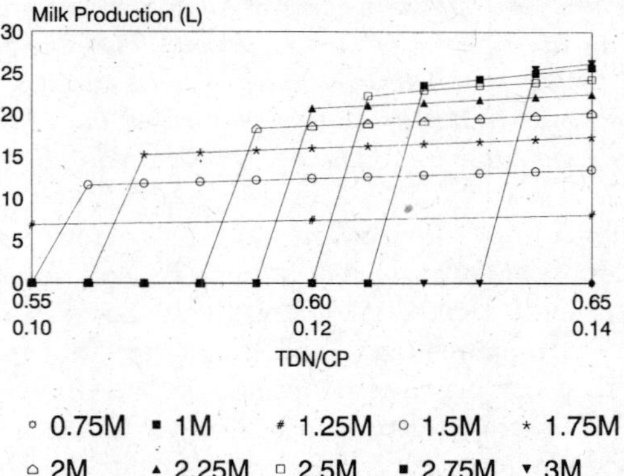

○ 0.75M   ▪ 1M   ⋆ 1.25M   ○ 1.5M   ⋆ 1.75M
△ 2M   ▲ 2.25M   □ 2.5M   ▪ 2.75M   ▾ 3M

Importantly, part of the increased need for external fossil fuels is caused by the decreasing availability of energy in, for example, forest reserves or biophysical capital (Giampietro et al., 1994). This tendency can be countered by reducing and/or adjusting effective demand, and/or more intelligent use of information (Chancellor, 1981).

Particularly LEIA systems adjust their effective demand, and they apply more elaborate knowledge / individual attention to their crops and animals. The farm systems in the New Conservation Agriculture (NCA) mode will mostly combine the use of more information with restricted use of external energy sources. The employment of both computers and indigenous technological knowledge (Warren, 1991) as well as the need for research, education and data systems are forms of induced innovations that attempt to meet this need. In this context it was interesting to read, at a late stage of the work, the following quote from Chancellor (1981):

imagine a garden-like situation with each worker managing a very small area [...]. On this basis, attention could be given to individual plants - weeds could be removed individually, water, nutrients, pest inspection and control means could be applied plant-by-plant. The persons could remember individual plants and their progress, harvest each one at the optimum time, apply remedial measures to poorly functioning plants or replace these with transplants as required. Soil could be tilled only in the seed zone [...].

He expresses that proper management uses information in addition to energy to achieve a higher output. However, the use of adjusted production objectives, system shape, i.e. criteria for system success is to be highlighted also. New Farm System Development will only defuse the rat race if it dares to propose drastically different options and criteria for development. Technology can only reduce energy dependency through clever use of information, together with the adjustment of system objectives: the challenge for NCA.

Maintenance requirements and criteria for system success

The coin of the low maintenance cost of simple (primitive) organisms has another side: at higher energy fluxes - the shape(s) of - these organisms are likely to perform less well than the more complex (developed) systems in terms of gross output, but not necessarily in terms of efficiency (Figures 2 and 3a-b). Though speculative, and without proof in this thesis, this might provide another key to the relation between system morphogenesis (drive and shape) and the usefulness of straw feeding methods, as discussed at the end of this chapter. But there is more, particularly relating with the criteria for system success (Ch. 2.2 and 5.1). Criteria are essentially an expression of the function, i.e. the demand or output from a system. Consequently, the goal setting of a system by the choice of criteria is bound to affect system morphogenesis (i.e. farm system evolution) through its influence on the resource / demand pattern. Again the importance of boundary conditions surfaces, and its relation with criteria setting deserves elaboration in the context of this thesis.

Table 1.   Energy efficiency and gross energy output from different systems

	Yield (kg/ha)	energy efficiency (kcal output/kcal input)
Mexico, Maize with manpower	1944	12.5
US, mechanized with horses	7000	3.4
US, mechanized with tractors	7000	2.4

Source: Tables 1, 2 and 3 in Pimentel (1984).

Figure 4   Energy ratios for food production (Source: Leach, 1976).

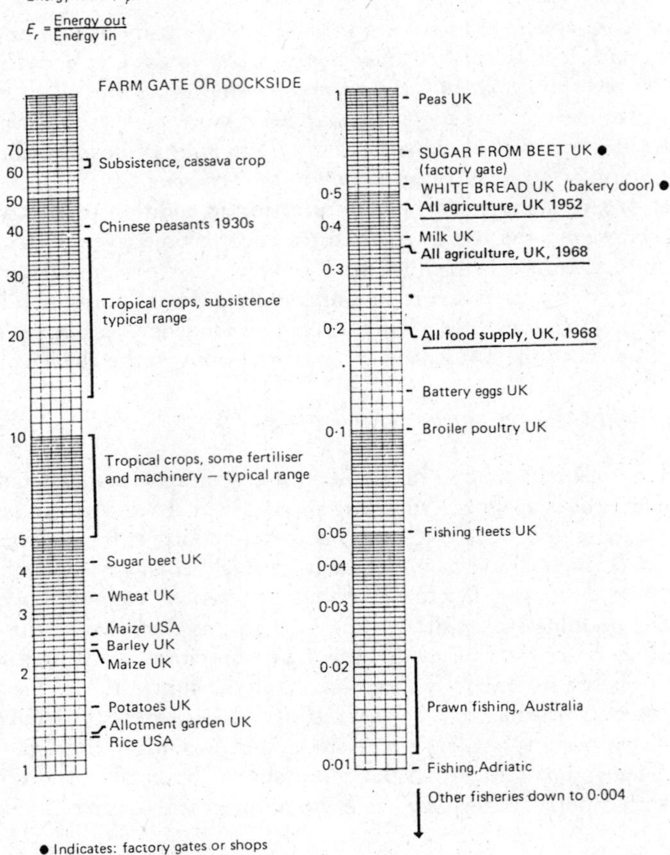

Firstly, it is generally accepted that systems tend to assume a shape that maximizes the utility of the limiting factor, whether land, labour or capital (Spedding, 1988). This implies that where straw and poor quality residues are plenty in comparison to concentrate supplements, it may be better to accept low individual animal output as a measure of system success. Obviously, this strategy, i.e. this shape, only works if the output of the system meets the effective demand, i.e. where low individual animal production can be compensated by large numbers of animals. As shown in Table 1 of Ch. 5.1, animal performance can be judged by at least three criteria: total feed conversion, concentrate conversion ratio and value conversion ratio. Importantly, each of these gives other optimal levels of animal output, i.e. each criteria of fitness 'selects' farm systems of another form. Clearly, the same holds true for crops, or for any other system, whether in fishery, manufacturing or education.

Secondly, and equally interesting in this context, it is clear that whereas one organism may perform better on the range of higher fluxes, it can be outcompeted at lower fluxes, and *vice versa*. See for example the difference between species A (the green algae) and species B (the blue algae) in Figure 2. According to this logic, and found true in a very general sense, both the *Bos indicus* cows and local grain varieties, might be expected to do better at lower fluxes, whereas the *Bos taurus* and high yielding varieties (HYVs) would be likely to perform better at higher fluxes. It is tempting to conclude that this genotype * environment interaction may be a reflection of underlying principles from system control. Whatever the truth, these principles show that the choice of optimal criteria for system success depends, therefore, on the range of conditions chosen, and criteria may need to be adjusted as conditions change (Ch. 2.2).

Thirdly, if import of energy or any other resource is required to achieve criteria of high output from one sub-system, a so-called.damning objective, one gets involved in the issue of trade-offs, particularly obvious in closed systems, at the expense of 'Simon' (see Ch. 2.1, 2.2 and 5.2). The concern about the effects of technology and development on equity (Conway, 1985) might well find its roots at least partly in this issue. It might provide an interesting clue to the answer to the question of Jackson (1981), i.e. whether a limited amount of good quality feeds should be allocated to a few elite herds, or equally spread over a much larger section of the animal population. As implied in figures 2Ap and 2Bp of Ch. 5.2, much depends on the feed quality of the basal feed in relation to the maintenance requirements of the (cow)subsystems. This point is also undoubtedly related to the issues raised by De Wit *et al.* (1987), i.e. whether agriculture should be concentrated on a few fertile areas or spread over larger, less fertile regions.

Last but not least, because criteria are to be different between systems (see also Behnke, 1985; Marten, 1988), it is logical for FSR practitioners to get involved in the specification of niche solutions and criteria. It is an administrators' nightmare (see quote A.J. de Boer in Ch. 2.2), but the administrators' perceived need for standard criteria might require improbable standardization, an energy intensive process, implying even extra waste. Policies that pursue standard criteria should be reconsidered, therefore, in the light of the tentative conclusion in Ch. 5.2 that desaggregation of feeds and animals tends to increase total system output. In other words, uniformity might optimize subsystem output, but not total (farm) system output. In fact, it may cause pollution by preventing the use of on-farm resources.

Efficiency versus total yield

The relation between the low maintenance cost for system control, on the one hand, and the low system output on the other hand, has yet another important consequence. In spite of a possibly higher energy efficiency of simple organisms at lower energy fluxes, their net and gross output can be too low to meet a given effective demand for food etc. by the organism or farm system itself, or by, for example, the urban population. This problem can be solved in expansion agriculture by allowing more animals to graze on the 'outfields', or by increasing individual animal output (Ch. 2.2). The expansion agriculture uses solar energy in plant biomass, the HEIA shifts its energy supply to fossil fuels, so-called energy subsidies (Odum, 1971), thereby tending to accumulate low flux resources. For both these modes of agriculture, this represents a case where the system has to cope with a damning objective, with associated problems of Simon effects, i.e. marginalization. The possibility to increase energy efficiency with better use of information should be specifically kept in mind (Chancellor, 1982; Bonny, 1993). Again, this can be seen the major challenge for the design of farm systems in NCA. The basic principles should include the use of adjusted demand, limited amounts of inputs, and diversity of subsystems, i.e. inclusion of 'scavengers' that can use 'waste' (= low flux resources).

CATTLE, STRAW AND SYSTEM CONTROL

The use of the second law of thermodynamics may seem far-fetched, but it is not new, and it provides essential information, i.e. a firm basis for an holistic approach to development and technology (Box 2). The relation between cattle, access to energy, wealth and system control of chapter 1 can now be better understood. As long as sufficient animals can be kept, they provide wealth, by converting solar energy that is available in biomass into products such as meat, milk, draught and speed. Each one of these products in one way or another, provides a form of control in the farm system or even society. Clearly, if the feed is of a better quality, i.e. of a higher energy flux, the systems that it supports can be more complex, potentially capable of yielding higher gross outputs. What is the relation with work on technology development for straw feeding systems?

Box 2:   A SUMMARY OF WORK ON THERMODYNAMICS ON (AGRICULTURAL) DEVELOPMENT.

The fact that thermodynamics and information theory can be useful beyond the physics of steam engines is known at least since the late sixties (Georgescu Roegen, 1971; Odum, 1971; Meadow, 1977; Johnson, 1981; Giampietro *et al.*, 1994), but folkwisdom knew it much longer, e.g. in variations on themes like 'nothing ventured, nothing gained' (Box 1 in Ch. 2.2).

Much work is done on energy analyses of systems (Leach, 1976; Pimentel and Pimentel, 1979; Grigg, 1982; Jones, 1989; Stout, 1990; Bayliss-Smith, 1991; Netting, 1993), providing a basis for thermodynamic analysis of agricultural development. Options to achieve savings of energy by using information are explained by Chancellor (1981), based among others on the relation between energy and information (Tribus and McIrvine, 1971). Also, some philosophers that studied the relation between development and society have touched on, or actually employed thermodynamic concepts (Illich, 1974; Commoner, quoted by Coolsaet, 1985; Ellul, 1990).

Technology, energy and information

Technology can be considered to combine the use of energy and information, a definition given in Ch. 2.2 that makes even more sense in this thermodynamic context. Proper use of information and energy can help to run a system cheaper in energetic terms, well worked out by Chancellor (1981). However, there is no technology that can make a system run without energy[2]. The use of energy and information, i.e. the money and resources spent on research about straw feeding systems, can help to make a system more energy efficient, for example, because:

---

[2] the first law states that energy cannot be created nor destroyed. Therefore the energy problem is caused more by the second law, that says that the form of energy tends to change, measured by the entropy production.

- once established rules (= information) from on-station trials, such as on maximum urea levels and minimum levels of water help to avoid waste of resources,
- the application of knowledge about optimum herd composition and animal nutrition from for example Ch. 5.2, or from various disciplines as in Biewinga *et al.* (1992) allows a form of New Farm Systems Development (NFSD) that get more from less.
- strategic use of information from on-farm trials, together with the design of a logical scale for farming systems development (Ch. 2.2) reduces the need for endless *ad hoc* experiments.
- rather than to emphasize the persuasion of farmers to adopt new technologies by using energy subsidies such as urea, polythene for straw treatment, it is necessary to emphasize the use of information on where and when to use these inputs in a most proper and informed way. This should be done according to the niche where the technology fits best, in spite of the administrators' quest for uniformity and blanket recommendations.
- urea treatment can make it possible to use a resource that otherwise might cause pollution problems, particularly if burning is the alternative. Ch. 4.1 elaborates some aspects of energy efficiency in the application of straw treatment with urea. It simplifies, however, by ignoring the possible use of straw for 'Simon', i.e. for mulch, fuel or other purposes.

At an abstract level, and in spite of obvious differences, straw feeding technologies are comparable with technologies such as a new bicycle frame, an airplane or a horse. Such analogies allow the use of information (experience) from other disciplines, and in spite of their risks, they help to put straw feeding methods in a broader context. Here it is interesting here to note the observation by Randhawa (1980) that refers to the use of horses in Aryan warfare:

[...] the domestication of the horse caused a great crisis in human history which may be compared to the invention of the steamship and later of the aeroplane in modern times. [...]
In the same manner, the introduction of fertilizer, high yielding varieties, milking robots or even the use of straw treatment can lead to great changes that resemble punctuated developments, not to mention the Simon effects, i.e. the social trade offs at other system levels. Ideally, in terms of - social - sustainability, what a technology could do is to run systems in a more energy efficient manner, or more properly in thermodynamic terms: with a minimum entropy production. It could allow leftover energy to remain with, or to be returned to 'Simon': making life of society as a whole easier, and allowing 'more with less', the hope (or illusion?) of WCED (1987). Practice is, however, that the owner of a technology tends to internalize the advantages of a technology, and to externalize the disadvantages.

Entropy, trade offs and common interest

If order in one subsystem requires more disorder elsewhere, it is tempting to attribute negative trade-offs from technology to increased total entropy (Rifkin, 1989; Ellul, 1990). Again, the use of analogy is dangerous, but the parallels are compelling. Besides the interest in this topic from ecologists and economists, philosophers have also worked on the topic (Box 2). Most of their work comes down to a critique on the illusion of progress by taking into account (= by internalizing) hidden costs, i.e. a true holistic system approach. For example:

an average car can be shown to transport its owner at a speed of approximately five (!) rather

than 80 kilometres an hour; if all costs are included, e.g. time required to earn the car, to
wash the car, to treat the accident victims, to build roads, or the time that pedestrians have
to wait to cross an ever more congested road (adapted from Illich, 1974).

These calculations may not be accurate, but they do reflect an awareness of the need for
an holistic approach. They again point to the side effects of - any - technology on equity:
a car, as well as a straw treatment technology can give the owner an edge over other
systems in the struggle for survival of the fittest individual. Indeed, damning objectives and
unintended Simon effects are common whenever the introduction of a technology in one
subsystem affects the boundary conditions of other subsystems. The running of subsystems
with damning objectives is likely to cause shortage of control elsewhere in the system,
particularly at the periphery, a problem that can go unnoticed or ignored in the centre for
quite some time (Weiskel, 1989; Kaplan, 1994). The lagtime in feed back, a tendency
present in all modes of agriculture, tends to increase oscillations in systems behaviour. It
will decrease sustainability if resistance to shock is a criterium for sustainability (Bok, 1964;
Meadows *et al.*, 1972; Conway and Barbier, 1990).

The opposite tendency to the fight for the survival of the fittest individual is found,
however, where (sub)systems use / design technology for common interest. This is a topic,
reflected in rules of many religions, and it should be of interest to disciplines such as
sociology, ecology, agricultural science and economics (Olson, 1971; Schumacher, 1973;
Johnson, 1981; Daly and Cobb, 1990; Bromley, 1992; Kraybill, 1993). The issue is a
particularly relevant topic to work out in the design of new mixed crop-livestock systems
that want to employ straw in their feeding schedules.

Communal ideotypes[3]

The principle of common interest is applied more frequently in crop- and livestock
production than may be realized. In cropping, for example, the relatively low yield of
densely planted individual plants results in a higher total yield of the entire plot (B.
Deinum, pers. comm. 1990). And Donald (1981) defines a communal ideotype for a grain
crop as follows:
[...] communal plants may give low individual plant yields, but when grown in a pure stand
at a density sufficient to induce interplant competition and full exploitation of the
environment, they are capable of high crop yields. It is proposed that any ideotype for wheat
or barley crops should be based on communal plants.
This thesis has referred several times to crop and/or livestock production systems where
low subsystem output is compensated by high total system output (Jones and Sandland,
1976; Nordblom, 1983; Kidane, 1984). Common interest is probably the core issue in the
design i.e. shape of new straw feeding systems, because exchange of resources between crops
and livestock requires mutual adjustment, almost by definition (Patil *et al.*, 1993; Ch. 5.2).

---

[3] The term communal here does not imply difference based on caste, religion or race as it does in some
cultures.

STRAW FEEDING AND SYSTEM MORPHOGENESIS

Finally, it is necessary to apply the concepts of energy flux and system morphogenesis to the usefulness of straw feeding methods in the system shapes of the central column of the classification matrix, as proposed in Ch. 2.3. Though tentatively, it appears possible indeed to discern at least some logic between the use of straw feeding methods and the drive and shape of farm systems. In fact, the two appear closely related, since the (straw) feeding system determines the system shape and vice versa. The exact relation with the drive of systems needs to be further elaborated, but the contours of these interrelations should be clear by now. The following discussion will review them, by proceeding from expansion agriculture, via LEIA towards HEIA, with New Conservation Agriculture (NCA) at the end. It intends to explain system evolution and/or morphogenesis, in relation with the use of induced innovations to cope with mankinds need for energy and information, to control society in one way or another. In the context of this thesis it appears appropriate to place NCA at the imaginary end point, well knowing that the sequence of farming system modes is not necessarily fixed (Ch. 2.2).

The mode of expansion agriculture is found where land, and therefore by implication feed, is not limited relative to the effective demand. Animals can eat the best feed available, selective consumption is done by the animal and encouraged by the farmer, in order to obtain high levels of energy intake and system control. Depending on the energy flux or the 'drive' in the feed, shape of the systems will tend to employ animals with lower or higher levels of production. Where (medium quality) forages are better available than high quality concentrate feeds, it can be more attractive to accept cheap liveweight gains (LWG) from a large herd than expensive LWG from a few high producers (Table 1, Ch. 5.1). Low individual production can thus be compensated by large herds, whether in traditional pastoral systems of Africa, or in modern ranching of Australia or the USA (Ch. 5.1). The best way to harness (solar) energy in those systems is to use animals that exploit a larger area than man can do alone by cropping. In those systems, animals truly are a form of power, and by implication: a form of wealth. Possession of animals can determine the control of men and women over systems, whether to wage war, to buy a bride or to impress the fellow citizens in any other way. Straw in those systems plays no significant role for mankind, since feeds with higher energy fluxes can be selected.

In Low External Input Agriculture (LEIA) the effective demand has increased relative to the access to land. As a consequence, and in absence of alternative sources of energy, the objectives of farming and consumption, i.e., the criteria for system success are adjusted to the resources. Roadside and rangeland grazing is limited, and even mere survival of animals - in the lean season - becomes a realistic objective for a farm family. Large scale fodder production is not possible and straw treatment cannot be afforded due to expenses such as for urea and polythene. Treatment of straws by using kitchen ash or animal urine might be a remote possibility, but there may not be enough straw to take full benefit of increased intake. Also, a temporary weight loss can be compensated by cheap gains on abundant roadside or grazing feed reserves in another season. Labour input is no problem in these systems, and the only way to ensure that all straw is eaten for herd survival, is to chop or soak the feed: selective consumption and wastage are prevented, weight loss is accepted as a means for survival. A system shape with low output in terms of milk or meat is the result.

High External Input Agriculture (HEIA) has no need for straw feeding since straw provides only a low energy flux, compared to fodders that are cultivated with, for example, fertilizers, irrigation and special cutting regimes. The rare exception is that limited quantities of straw in rations of high producing cows of these systems can serve to maintain rumen function on diets with low fibre. But this does not invalidate the point of the energy flux: straws in those rations are primarily fed to provide fibre, and not to provide energy to the animal. The high output of individual animals in these HEIA systems is accompanied by high requirements for system maintenance in the form of housing, management, veterinary care, feed supply and market infrastructure, both permitting and necessitating high total output. A major problem in HEIA systems is, however, the disposal (=externalization) of their waste products. It is no wonder that this - relatively recent - boundary condition of disposal problems is forcing these systems towards recycling, a morphogenesis towards NCA. It is a challenging area of work, both for those obsessed by technology, as for those who prefer to adjust demand patterns. If and when systems move into NCA, the criteria for cow subsystem success may have to be adjusted. Cows with medium production levels may have to be reintroduced, they may regain status as waste converters, or seasonal production patterns may turn out to be useful after all, obviously not without trade-offs (Ch. 2.2 and 5.2).

New Conservation Agriculture (NCA) has some access to outside sources of energy, combined with adjusted system objectives, flexible criteria for subsystem success, and maximum emphasis on use of information, i.e. management. The limited energy flux in NCA can be either due to still limited but increasing access to energy in systems originating from LEIA, or to waste disposal problems in systems 'regressing' from HEIA. This intermediate energy flux in NCA provides conditions, where straw can profitably be fed as such or treated with urea. This tallies with the observations in Ch. 4.1, that straw treatment is mainly useful in systems between high and low input, i.e. where cultivated fodder and concentrate feeds are scarce, and where the production of the animal is adjusted to the available feed. Some access to the market is, however, necessary to purchase the external inputs, a precondition that is fulfilled in this mode of agriculture. NCA has, however, additional features in terms of options for feed supply based on a wider range of crop residues than usually taken into account. A variety of trees, fodders and catchcrops, are utilized in this mode of agriculture, in order to protect land from erosion, or to prevent leaching of mineralized nutrients in fallow periods, or to recycle leached nutrients from the subsoil. Each of these can be called crop residues, and an important additional class of these residues are those of grain milling and oilseed processing. In combination with these, interesting optimization problems for the use of straw occur, including the need to use straw as bedding for mulch.

EPILOGUE

It costs something to get something: generalization goes at the expense of detail (Traub and Wozniakowski, 1994). Generally speaking, high and unlikely forms of production require more and better resources in terms of energy and information. Conversely, low energy and information flux resources do not allow highly developed systems. Consequently, the shape of systems appears to change together with the drive, here called the flux in the system. Nevertheless, straw and grass represent forms of energy, and they are, therefore, potentially useful to help organizing and maintaining a farm system and a society. It is, therefore, no surprise that animals - as converters of this energy - are considered a source of wealth and power in societies with sufficient access to feed where more and/or higher producing animals permit higher levels of system control. Straws are feeds with a lower energy flux than concentrates and good grasses, and, therefore, the systems based on straw feeding will take their shape accordingly, as will the role of cattle as a factor in system control.

This thesis does not pretend to provide detailed answers to specific conditions but it has helped:
- to explain the usefulness of straw feeding methods in different farming systems, based on field observations, common sense, results from component research, economic calculations and confirmed by theories about system control,
- to avoid a large amount of *ad hoc* experiments on the use of straw feeding methods or any other technology; it appears possible instead, to form hypotheses about the applicability of technology before proceeding to on-farm trials and extension programs, saving resources that can be used elsewhere,
- to develop a framework for further work by providing a set of scales that appear to assist in the explanation of system morphogenesis; i.e. the drive and shape of systems,
- to relate the role of straw feeding methods with the drive and shape of cow-, farm- and society systems
- to show that an holistic approach requires an interdisciplinary approach, and flexible and changing criteria for (sub)system development,
- to provide evidence for the fact that subsystem output may have to be adjusted to the priorities and boundary conditions of the overall system.

Undoubtedly, but beyond this thesis, more clarity and new questions about straw feeding systems can be found by combining energy and information from other disciplines than only animal nutrition. Archetypically for farming systems research, the use of concepts from other disciplines (knowledge and information systems), such as ecology, thermodynamics and social sciences, is likely to be useful, also beyond these issues of cattle, straw and systems control.

ACKNOWLEDGEMENTS

Special thanks are due to W. Van Densen who introduced me to the concept of system control on a scale from fisheries to aquaculture, to J. Lyklema who took time to teach me about entropy, to H.P.P. Kessels who supplied me with some very useful references, to P.F.M. Elenbaas who helped to define dissipative structures and who explained the difference between blue and green algae, and to M. Harding of the National Museum of Science and Industry in London for a brief but inspiring discussion about perpetual motion.

REFERENCES

Achterhuis, H., 1975. Filosofen van de Derde Wereld: Frantz Fanon, Che Guevara, Paolo Freire, Ivan Illich, Mao Tse-Toeng. Bilthoven, The Netherlands. 127 pp.
Bayliss Smith, T.P., 1991. The Ecology of Agricultural Systems. Cambridge University Press, 112 pp.
Behnke, R.H., 1985. Measuring the benefits of subsistence versus commercial livestock production in Africa. Agric. Syst., (1985): 109-135.

Biewinga, E.E., Aarts, H.F.M., and Donker, R.A., 1992. Melkveehouderij bij stringente normen, bedrijfs- en onderzoeksplan van het Proefbedrijf voor Melkveehouderij en Milieu. September 1992, De Marke, Hengelo, Rapport no.1.

Bok, S.T., 1964. Cybernetica (stuurkunde): Hoe sturen wij ons leven, ons werk en onze machines? Aula, Antwerpen, 259 pp.

Bonny, S., 1993. Is Agriculture Using More and More Energy? A French Case Study. Agric. Syst., 43: 51-66.

Bromley, D. (Editor), 1992. Making the Commons Work: Theory, Practice and Policy. Institute for Contemporary Studies. San Francisco. 339 pp.

Cavalli-Sforza, L. and Feldman M., 1981, Cultural Transmission and Evolution, Princeton University Press, Princeton, N.J., USA

Chancellor, 1981. Substituting information for energy in agriculture. Transactions of the ASAE, (1981): 802-807.

Coolsaet, W., 1985. De verkeerde wereld - opstellen over Illich, Commoner, Toffler, Kuhn en Feyerabend, Althusser: Acco, Leuven / Amersfoort, 225 pp.

Conway, G.R., 1985. Agroecosystem Analysis. Agric. Admin., 20: 31-55.

Conway, G.R., and Barbier, E.B., 1990. After the Green Revolution, Sustainable Agriculture for Development, Earthscan Publications, London, 205 pp.

Cox, G.W. and Atkins, M.D., 1974. Agricultural Ecology, an Analysis of World Food Production Systems. W.H. Freeman and Company, San Francisco.

Crosby, A.W., 1986. Ecological Imperialism, the Biological Expansion of Europe (900-1900). Cambridge University Press. Cambridge, U.K.

Daly, H.E. and Cobb, J.B., 1990. For the Common Good: Redirecting the Economy towards Community, the Environment and a Sustainable Future. Green Print, 482 pp.

Darwin, C., 1859. The Origin of Species by Means of Natural Selection. John Murray.

Dawkins, R., 1982. The extended phenotype, the long reach of the gene. Oxford University Press, Oxford / New York, 307 pp.

Dawkins, R., 1991. The Blind Watchmaker. Penguin, London. 340 pp.

De Wit, C.T., Huisman, H. and Rabbinge, R., 1987. Agriculture and its environment: are there other ways? Agric. Systems, 23(3): 211-236.

Dieterich, F.G., 1986. Dieterich's 50 Perpetual Motions. Lindsay Publications, POBox 12, Bradley IL 60915, USA.

Donald, C.M., 1981. Competitive plants, communal plants, and yield in wheat crops. p. 223-247. In: L.T. Evans and W.J. Peacock (Editors), Wheat Science - Today and Tomorrow.

Eiseley, L., 1957. The Immense Journey; An Imaginative Naturalist Explores the Mysteries of Man and Nature. Vintage Books, USA. 211 pp.

Elenbaas, P.F.M., 1994. Field and Labwork on Ecology in Zimbabwe, Centre for Development Cooperation Services. Vrije Universiteit, Amsterdam, The Netherlands. 169 pp.

Ellul, J., 1990. The Technological Bluff. Translated by G.W. Bromley. William B. Eerdmans Publishing, Grand Rapids Michigan. 418 pp.

Frisch, J.E. and Vercoe, J.E., 1978. Utilizing breed differences in growth of cattle in the tropics. World Anim. Rev., 25: 8-12.

Giampietro, M., Cerretelli, G. and Pimentel, D., 1992. Energy analysis of agricultural ecosystem management: human return and sustainability. Agriculture, Ecosystems and Environment, 38: 219-244.

Giampietro, M., Bukkens, S.G.F. and Pimentel, D., 1994. Models of energy analysis to assess the performance of food systems. Agric. Syst., 45: 19-41.

Georgescu-Roegen, N., 1974. The Entropy Law and the Economic Process. Harvard University Press, Cambridge, Massachusetts. 457 pp.

Gleick, J., 1987. Chaos, Making a New Science. Abacus Book, London, 352 pp.

Grigg, D., 1982. The Dynamics of Agricultural Change, the Historical Experience. Hutchinson, London, 260 pp.

Hayami, Y. and Ruttan, V.W., 1985. Agricultural Development, An International Perspective. The John Hopkins University Press, Baltimore and London.

Illich, I., 1974. Energy and Equity. Calder & Boyars, London and Harper & Row, Publishers. Inc. New York. 84 pp.

Jackson, M.G., 1981. A new livestock development strategy for India. World Anim. Rev., 37: 2-8.

Johnson, L., 1981. The thermodynamic origin of ecosystems. Can. J. Fish. Aquat. Sci., 38: 571-590.

Jones, M.R. 1989, Analysis of the use of energy in agriculture - approaches and problems, Agricultural Systems, 29, 339-355

Jones, R.J. and Sandland, R.L., 1974. The relation between animal gain and stocking rate, derivation of the relation from the results of grazing trials. J. Agric. Sci., Camb.83: 335-342.

Kaplan, R.D., 1994. The coming anarchy. The Atlantic Monthly, 2(1994): 44-76.

Kessels, H.P.P., Giesen, G.W.J., Lotan Singh, Harika, A.S. and Schiere, J.B., 1994. Energy for Sustainable Crop-Livestock Production: Waste or Treasure? Paper presented at the third International Livestock Farming Systems Symposium, held at September 1-3, 1994, at MLURI, Aberdeen, UK.

Kidane, H., 1984. The Economics of Farming Systems in the Different Ecological Zones of Embu District (Kenya), with Special Reference to Dairy Production. Ph.D. Thesis, Hannover University, Hannover, 227 pp.

Kiran Singh and Schiere, J.B. (Editors.), 1993. Feeding of Ruminants on Fibrous Crop Residues: Aspects of Treatment, Feeding, Nutrient Evaluation, Research and Extension. Proc. Int. Workshop, 4-8 february 1991, NDRI-Karnal, ICAR, New Delhi, 486 pp.

Kramer, N.J.T.A. and De Smit J., 1987. Systeemdenken, 4e herziene druk, Stenfert Kroese, Leiden/Antwerpen[4]

Kraybill, D.B., 1993. The Riddle of Amish Culture. The Johns Hopkins University Press, Baltimore and London.

Leach, G., 1976. Energy and Food Production. International Institute for Environment and Development, London, 151 pp.

Lewin, R., 1993. Complexity: Life on the Edge of Chaos. Phoenix, London, 208 pp.

Lyklema, J., 1991. Fundamentals of Interface and Colloid Science, Vol. 1: Fundamentals. Academic Press, Great Yarmouth, Norfolk.

Marten, G.G., 1988. Productivity, Stability, Sustainability, Equitability and Autonomy as Properties for Agroecosystem Assessment. Agric. Syst., 26: 291-316.

Maynard Smith, J. (ed.), 1982. Evolution now, a century after Darwin. Nature, 239 pp.

Maynard Smith, J., 1989. Evolutionary Genetics, Oxford University Press, Oxford / New York / Tokyo, 325 pp.

Meadows, D.H., Meadows, D.L. and Randers, J., 1972. The Limits to Growth: A Report for the Club of Rome's Project on the Predicament of Mankind. New York, 207 pp.

Meadows, D.L. (Editor), 1977. Alternatives to Growth I: A Search for Sustainable Futures. Ballinger, Cambridge, Massachusetts. 393 pp.

Netting, R.McC. 1993. Smallholders, Householders, Farm Families and the Ecology of Intensive, Sustainable Agriculture. Stanford University Press, Stanford.

Nordblom, T.L., 1983. Livestock-Crop Interactions, the Decision to Harvest or to Graze Mature Grain Crops. Discussion paper No.10, Inter. Center for Agric. Res. in the Dry Areas (ICARDA). 21 pp.

Odum, H.T., 1971. Environment, Power and Society. Whiley Interscience, New York, 331 pp.

Odum, H.T., 1983. Systems ecology, an introduction. Whiley Interscience, New York. 644 pp.

Ogink, N.W.M., 1993. Genetic size and growth in goats. PhD-thesis, Agricultural University Wageningen, The Netherlands.

Olson, M., 1971. The logic of collective action, Public Goods and the Theory of Groups, Harvard University Press, Cambridge/Massachusetts/London/England. 186 pp.

Patil, B.R., Rangnekar, D.V. and Schiere J.B., 1993. Modelling of crop-livestock integration: effect of choice of animals on cropping pattern. In: Kiran Singh and J.B. Schiere (Editors), Feeding of Ruminants on Fibrous Crop Residues: Aspects of Treatment, Feeding, Nutrient Evaluation, Research and Extension. Proc. Int. Workshop, 4-8 february 1991, NDRI-Karnal. ICAR, New Delhi, pp. 336-343.

Pimentel, D. and Pimentel, M., 1979. Food, Energy and Society. John Wiley, New York.

Pimentel, D., 1984. Energy flow in agroecosystems. In: R. Lowrance, B.R. Stinner and G.J. House, Agricultural Ecosystems, Unifying Concepts. Whiley Interscience, John Wiley, New York / Chichester / Brisbane / Toronto / Singapore.

---

[4] as if to highlight the importance of information, a part of the most relevant chapter in this book was missing from the library copy of our university. At the time of completion of the proofs there was no time left to look for a more complete version, but the remaining parts strongly suggest that this reference is useful (for those who understand Dutch information!)

Ponting, C., 1991. A Green History of the World, Penguin Books, London. 432 pp.

Prigogine I. and Stengers I., 1985. Order out of Chaos, Man's New Dialogue with Nature. Flamingo, London, 349 pp.

Randhawa, M.S., 1980. A History of Agriculture in India. Volume I. Beginning to 12th Century. Indian Council of Agricultural Research, New Delhi. 541 pp.

Reiss, M.J., 1989. The Allometry of Growth and Reproduction. Cambridge University Press, Cambridge.

Rifkin, J., 1989. Entropy into the Greenhouse World. Bantam Books, New York, revised edn., 354 pp.

Schumacher, E.F., 1973. Small is beautiful. Abacus, Sphere books Ltd., Penguin, London, UK. 255 pp.

Shaner, W.W., Philipps, P.F. and Schmehl, W.R., 1982. Farming Systems Research and Development: A Guideline for Developing Countries. Westview Press, Boulder, Colorado.

Shaner, W.W., Philipps, P.F. and Schmehl, W.R., 1982. Readings in Farming Systems Research and Development. Westview Special Studies in Agriculture / Aquaculture Science and Policy.

Simmonds, N.W., 1986. A short review of farming systems research in the tropics. Exp.Agric. Vol. 22: 1-13.

Simmons, I.G., 1989. Changing the Face of the Earth, Culture, Environment, History. Basil Blackwell, Oxford / New York, 487 pp.

Spedding, C.R.W., 1988. An introduction to Agriculture Systems, 2nd Edition, Elsevier, London.

Stout, B.A., 1990. Handbook of Energy for World Agriculture. Elsevier, London. 504 pp.

Traub, J.F. and Wozniakowski, H., 1994. Breaking Intractability, Scientific American, January: 90-93.

Tribus, M. and McIrvine, E.C., 1971. Energy and information: information controls energy flows, and the two are also related at a deeper level. Scientific American, 224/225[5](3): 179-188.

Turpin, D.H., 1988. Physiological Mechanisms in Phytoplankton resource competition, Ch. 8, p. 316-368. In: Sandgren C.D. (Editor), Growth and reproductive strategies of freshwater phytoplankton, Cambridge University Press, Cambridge. 442 pp.

Warren, M.D., 1991, Using Indigenous Knowledge in Agricultural Development, World Bank Discussion Papers nr. 127, Washington, USA, 46 pp.

WCED, 1987. Our Common Future. World Commission on Environment and Development, Oxford University Press, Oxford, 400pp.

Weiskel, T.C., 1989. The ecological lessons of the past - an anthropology of environmental decline. The Ecologist 19(3): 98-103.

Westgaard, P. and Sundstøl, F., 1986. History of straw treatment in Europe. pp.155-163. In: M.N.M. Ibrahim and J.B. Schiere (Editors), 1986. Rice Straw and Related Feeds in Ruminant Rations. Proc. Int. Workshop, Kandy, Sri Lanka, SUP Publication No.2. 407 pp.

Wilkinson, R.G., 1973. Poverty and Progress: An ecological Model of Economic Development. Methuen, London.

Zevenboom, W., 1986. Ecophysiology of Nutrient Uptake, Photosynthesis and Growth. Canadian Bulletin of Fisheries and Aquatic Sciences, 214: 391-422, Special issue on Photosynthetic Picoplankton, T.Platt and W.K.W. Li (eds.).

---

[5] through another remarkable quirk of fate (see footnote 4), the Volume Number is misprinted on this particular issue of the Scientific American. Those who would look in their library for No. 224 would soon find out the importance of information in system control: the proper volume number should be 225, though the issue indicates that it is nr. 224.

SUMMARY

SAMENVATTING

CURRICULUM VITAE

'Ach, voor mijn part vergeet je alles,
als je een ding maar onthoudt
dat je heel veel weg moet gooien,
voor je echt iets overhoudt'
        Herman van Veen

'an event is something that happens at a particular point in space and at a particular time. So
one can specify it by four numbers of coordinates. Again, the choice of coordinates is arbitrary;
one can use any three well-defined spatial coordinates and any measure of time. In relativety,
there is no real distinction between the space and time coordinates, just as there is no real
difference between any two space coordinates'.
        Hawkins, S.W., 1988, A brief history of time, from the big bang to black holes, Bantam Books, Toronto
        / New York / London / Sydney / Auckland

SUMMARY

Crop residues are an important feed resource for livestock in many temperate and tropical farming systems. These residues are generally understood to consist of straws, and of by-products from grain milling and oilseed processing. However, they can also include leaves and fodder from trees, grasses, green manures and / or catchcrops which are grown to support cropping through soil conservation measures. Straws are often considered to be poor quality feeds, but their widespread use calls for a further analysis of their importance and methods of feeding. This thesis, therefore, reports a form of system analysis that aims to understand two related questions. The first concerns *the suitability of straw feeding methods in different farming systems*, and the second addresses *the role of straws in the drive and shape of farming systems*. The study of, and answers to these questions are described in four major sections, preceded by an introduction and followed by a discussion.

Cattle and Straw

The introduction starts by relating the value of livestock feed to the value of animals and their products (Ch. 1). It sets a theme of the thesis i.e. that animals traditionally represent a form of wealth, capital and money. A major reason for this is proposed to be that animals convert solar energy that is captured in plants (=feed biomass) and not suitable for human consumption, into products that are useful to mankind. When the feed supply changes, the role of animals and the method of feeding are bound to change (Ch. 2.2, 2.3 and 6). In many farming systems straw feeding becomes increasingly important, for example because increased demand for cropland leaves less land for grazing. As a result, much research has been done on technical aspects of different straw feeding methods. These methods include chemical, physical and biological treatments, supplementation of straws with better feeds, and use of agronomic measures to improve the quality or quantity of straw.

Ammonia treatment of straw for cattle feed, alone or in combination with supplements, has caught special attention of researchers and policy makers in many parts of the world. The method is technically feasible under farmers conditions, based on the use of ammonia in temperate regions, and on the use of urea in the tropics. Besides a refinement of the technical aspects, however, there was need for a more comprehensive understanding of the usefulness of these and other feeding methods, in a range of farming systems: the topic of this thesis.

Farming Systems Research

The study of straws in different farming systems is based on a review of approaches, definitions, history and different forms of Farming Systems Research (FSR) (Ch. 2.1). It discusses important concepts and methodologies, such as agroeco-zoning, similarities between cropping and livestock systems research, the occurrence of trade-offs between system objectives, and the use of thought experiments as a form of modelling to explore system behaviour. The review justifies the need for further study of farming systems, and it identifies fundamental issues in FSR, e.g. the importance of the definition of system boundaries. It also observes that all systems appear to be subject to the same basic rules, whether these systems are living cells, cows, farm systems or societies.

System classification and development

Rather than to undertake an in-depth study of the use of straw in one system, the work reported in this thesis aimed to understand the relation between straw feeding methods and system development over a large range of situations. To do so, it was necessary:
a)   to identify a large variety of farming systems, obtained from literature and personal observations,
b)   to arrange them into a classification that explains system behaviour.
This exercise resulted in a classification that uses a two dimensional matrix, of which the vertical axis expresses the relative scarcity of classical production factors: land, labour and capital in relation to the demand for (animal) products. The horizontal axis represents a scale of systems with either predominantly livestock, crops, or a mix of both (Ch. 2.2). The combination of relative access to production factors with demand for animal products is called the resource / demand pattern, and it is hypothesized to determine system development. In line with that assumption, Ch. 2.3 indicates the usefulness of straw feeding systems per subclass of the classification matrix.

As the work progressed, it became increasingly clear that system development can be equated with system morphogenesis, and that resource / demand patterns can be equated with boundary conditions. Both these terminologies are used in the more fundamental discussion of system development in Ch. 6. As such, the classification based on resource / demand patterns replaces the traditional distinction between tropical and temperate systems with the distinction between open and closed systems (Ch. 5.1). This is indeed, a more relevant and pertinent classification for the purposes of this thesis.

The discussion in Ch. 2.2 also shows that development of a subsystem does not always lead to an overall improvement. Development is often a response to changes in relative scarcities: *necessity is the mother of invention*. In practical terms, necessity concerns the need for commodities and services such as food, clothing, transport, pleasure and housing. At a more abstract level, necessity concerns the need for energy in various guises. Importantly, the effective demand is the result of the product of the number of people and per capita consumption, it is not only determined by the size of the population! Necessity forces the system to adopt and search for so-called 'induced innovations'. The search for, and application of new straw feeding methods is a typical example of such an induced innovation. Just as with the introduction of many other technologies, these innovations may mean progress for one system, but they often appear to be associated with the use of resources from other systems. In this context, Ch. 2.2. introduces the so-called Simon effect, representing the principle that a higher output achieved in one system is often achieved at the expense of the output in another system.

Component research on straw feeding methods

In order to predict the usefulness of some straw feeding methods, particularly those associated with ammonia treated straw, a series of animal feeding trials was done. They compared the use of treated straw with that of untreated straw, alone or combined with supplements (Ch. 3.1-3.4). To make the data more generally applicable, the trials were increasingly designed to determine basic animal nutrition parameters, i.e. feed digestibility and intake. The work in Ch. 6 showed later that these parameters actually represent the energy flux in systems. The feeding trials, together with information from literature, show that urea ammonia treatment consistently increases straw intake and the availability of digestible energy from the straw for the animals. The crude protein content of the straw also increases after urea-ammonia treatment, but that effect can be easier achieved with supplementation of the straw rations. The use of small amounts of supplements, to achieve so-called catalytic supplementation, can also increase the digestibility and intake of straw, but the response appears to be smaller and less consistent than with treatment. This method may only be relevant at specific ranges of supplementation: particularly around maintenance levels of production.

Usefulness of straw treatment

The economic suitability of straw treatment was determined with thought experiments, i.e. economic calculations that used parameters obtained from literature (Ch. 4.1). Testing of these results was done with nutritional and socio-economic considerations (Ch. 4.2 and 5.1), and data from the animal nutrition trials (Ch. 3.1 - 3.4). This confirmed that straw treatment, such as other feeding methods, is a typical 'niche' innovation. In other words, the answer to the first question of this thesis is that the usefulness of straw feeding systems is highly system specific. It is shown for example, that straw treatment can be said to be useful mainly:
- where plenty of straw is available, i.e. where straw is cheap compared to other feeds,
- where high individual animal production is not required,
- where the farm system has boundary conditions that allow access to inputs in exchange for outputs from the animals.

With the benefit of hindsight, it is not surprising that ecological terminology such as niche solutions and system morphogenesis is used. Particularly during the final stage of the work on this thesis, it became apparent that farm system development resembles an evolution. In short, it resembles Darwin's *origin of species by the survival of the fittest*, with a Lamarckian twist, i.e. *induced innovations* become *acquired characters*. This agrees with the reasoning in Ch. 2.2 where resource / demand patterns, i.e., the boundary conditions combined with the use of innovations, are suggested to determine farm system development.

Open and closed systems, damning objectives and the Simon effect

Further attention to the importance of boundary conditions for the usefulness of (straw)feeding systems was warranted. The distinction into closed and open systems as identified in Ch. 2.2 was, therefore, elaborated in Section 5. The behaviour of low external input agriculture (LEIA), is compared with that of high external input agriculture (HEIA)

(Ch. 5.1). A major difference appears to be that the 'closed' systems in LEIA tend to adjust their objectives to the resources, whereas the 'open' systems of HEIA tend to adjust the resources to their objectives. This tentative conclusion led to a new set of thought experiments which assumed conditions of closed systems. Unlike in the commonly 'open' systems approaches of agricultural planning, the access to (feed) resources was assumed to be fixed.

The thought experiments explored what happens if no feed can be purchased from outside the system (Ch. 5.2). The results showed that in those 'closed' system conditions, production targets for subsystems may need to be adjusted to the resources, in order to extract the maximum system output. If a production target of a subsystem is set too rigid and too high for the subsystems' resources, it becomes what is here called a damning objective. This term implies a target that leads to either an unfeasible solution, or to the need to extract the lacking resources from other systems. A damning objective exhausts the system, it leads to equity problems such as expressed in the Simon effect, and it can even prevent the use of low quality resources available on-farm. Hence, a resource such as straw becomes a waste and pollution is the result. In addition, the thought experiments provided evidence that in closed systems, the total system output in, for example, terms of milk, dung and meat can be increased with a combination of low and high producers, rather than with one standard cow or farm system. This principle is likely to apply for any system, whether at the level of a cow, a farm, or a region. It implies that the use of one type of (cow)system, i.e. one uniform criterium for (sub)system output, may reduce rather than maximize the overall system output.

The drive and shape of systems

The discussion in Ch. 6 ties it all together, and it also provides a tentative answer to the second question of this thesis. It reviews and verifies the conclusions obtained thusfar, and it appears to provide clues that explain loose ends of the thesis, such as the need for changing and different criteria, the Simon effect and equity problems in development. The use of concepts from thermodynamics and information theory appears to be crucial. They explain the central role of energy and information in the control of systems. They also appear to confirm the idea that technology can be defined as a combination of inputs (energy) and management (information), as given in Ch. 2.2.

The second law of thermodynamics could provide the key to the question on the drive and shape of systems. It states that all systems when left on their own, tend to a state of maximum entropy, i.e. lack of order. This explains, at least in part, the drive of systems. As a consequence, both energy and information are needed to organize, maintain or control a system, again regardless of whether it concerns a living cell, an animal, a farm or society. Based on these concepts, it appears possible to discern a relation between the suitability of feeding systems, and the drive and shape of systems over a range of resource / demand patterns. As the nutritive quality of feeds increases in terms of digestible energy fluxes, it is possible to maintain more complicated systems, that have a generally higher absolute output: the relation between energy density in feed and shape of systems is clear. In that sense it should be no surprise that treated straw is too good for animals around maintenance, but useful for animals with a medium level of production. Not only the fact that treated straw is useful for systems with medium production levels is relevant, there is more. If feeds such as treated straw are the only ones available, they permit and thereby shape only medium levels of production. On a more general level, it is likely that according to their energy fluxes, feeds or any other energy source, drive and shape systems ranging from simple or primitive, to complex or developed. Animals, by converting the energy in feeds, are thus one way to help organize and maintain society, which depending on the type of feed and animals, is permitted to become developed or to remain primitive. Animals in that sense, if properly kept and fed, clearly imply a source of wealth and system control. The use of information from other disciplines (knowledge systems) is a typical prerequisite for Farming Systems Research. It is also likely to further discern or even quantify the logic behind issues such as the role of cattle and straw in system control.

SAMENVATTING

Vezelrijke gewasresten (zoals stro), zijn een belangrijk veevoer in veel bedrijfsstelsels zowel in de gematigde als de tropische klimaatstreken. Stro vormt hierbij echter slechts een deel van een veel grotere groep gewasresten, waartoe b.v. ook de bijprodukten van de graan- en oliezaadverwerking behoren. Zelfs bladeren van bomen, en voer van gras en andere gewassen, die men plant ter bescherming van de bodem, kunnen beschouwd worden als gewasresten. Stro wordt vaak gezien als slecht veevoer. Echter, het veelvuldige gebruik ervan als zodanig vraagt om nader onderzoek, met name naar het belang van de diverse strovoedermethoden in de verschillende of veranderende bedrijfsstelsels. In dit proefschrift worden d.m.v. een vorm van systeemanalyse twee met elkaar samenhangende vragen beantwoord. De eerste vraag betreft de *geschiktheid van strovoedermethoden in verschillende bedrijfsstelsels* en de tweede richt zich op *de rol van stro in de ontwikkeling en vorming van bedrijfsstelsels*. Het proefschrift bestaat uit vier hoofdonderdelen, voorafgegaan door een inleiding en gevolgd door een discussie.

Vee en Stro

In de inleiding wordt het verband aangegeven tussen de waarde van veevoer en het sociaal-economische belang van dierlijke produktie (Hfdst. 1). Het hebben van vee vertegenwoordigt immers traditioneel een vorm van rijkdom, status en macht. Een belangrijke reden hiervoor lijkt te zijn, dat plantaardig materiaal door dieren omgezet kan worden in voor de mens nuttige produkten. Een achterliggende verklaring is, dat vee een voor mensen niet direkt te gebruiken vorm van energie om kan zetten in een wel direkt bruikbare vorm. Bij een zich wijzigend aanbod van veevoer (als vorm van energie) en het beschikbaar komen van andere, veelal op fossiele energie gebaseerde produktiemiddelen, zal dan ook mettertijd de rol van vee en de manier van voeren veranderen. Ondanks haar lage voederwaarde, wordt stro in veel bedrijfssystemen over de hele wereld echter een steeds belangrijker voer o.a. omdat door een grotere vraag naar akkerbouwprodukten de beschikbaarheid van graas- en grasland afneemt.

Sinds ca. 1900 is er door officiële onderzoeksinstellingen veel onderzoek gedaan naar de technische aspecten van verschillende methoden van strovoedering. Zij omvatten chemische, fysische en biologische ontsluiting naast bijvoedering met betere voeders. Recentelijk is er ook meer belangstelling gekomen voor het gebruik van teeltmaatregelen om de kwaliteit en opbrengst van stro te verbeteren. Chemische ontsluiting van stro met ammoniak, zowel met als zonder bijvoedering, heeft de speciale aandacht getrokken van onderzoekers en beleidsmakers in veel gebieden van de wereld. De methode wordt op beperkte schaal toegepast op zowel grote als kleine bedrijven. In tropische klimaten wordt het stro ontsloten met ureum en onder gematigde klimaatomstandigheden met ammoniak. Naast een verfijning van de technische aspecten, was er echter behoefte aan een beter begrip van het nut en de rol van deze en andere strovoedermethoden in verschillende bedrijfssystemen; de direkte aanleiding tot het schrijven van dit proefschrift.

Onderzoek aan landbouwstelsels

Het onderzoek naar de rol van stro in verschillende landbouwstelsels wordt begonnen met een overzicht van de geschiedenis, benadering en soorten van landbouwstelselonderzoek. Deze tak van onderzoek wordt in het engels Farming Systems Research (FSR) genoemd (Hfdst. 2.1) en het kan zowel academisch als praktisch gericht zijn. Het overzicht verschaft belangrijke begrippen en definities, zoals verschillende interpretaties van het woord systeem, overeenkomsten tussen gewas- en veeteeltstelselsonderzoek en het maken van afwegingen van voor- en nadelen (de trade-offs) van de systeemdoelen. Daarnaast wordt het gebruik van denkexperimenten als een simpele vorm van modelleren met het doel systeemgedrag te verkennen, uitgelegd. Het overzicht rechtvaardigt verder de noodzaak om door te gaan met FSR, maar het toont ook enkele fundamentele problemen in FSR, zoals de definitie van systeemgrenzen. Uit het overzicht blijkt verder dat er vele analogieën bestaan tussen het gedrag van op het oog verschillende systemen, zoals een cel, een koe, een landbouwstelsel of een samenleving.

Indeling en ontwikkeling van systemen

De eerste vraagstelling van dit proefschrift betreft het nut van strovoederingsmethoden in verschillende bedrijfsstelsels. Indirekt betekent dit, het zoeken naar de verklaring van de relatie tussen strovoederingsmethoden en de ontwikkeling van landbouwstelsels. Daarvoor was het nodig om:
a)   een groot aantal landbouwstelsels te beschrijven, op basis van literatuurgegevens en eigen waarneming
b)   deze landbouwstelsels op een zodanige manier te rangschikken, dat systeemgedrag kan worden verklaard.

Deze benadering leverde een indeling op, die gebruik maakt van een tweedimensionale matrix. De verticale as daarvan geeft de relatieve schaarste aan klassieke produktiefaktoren: land, arbeid en kapitaal in relatie tot de vraag naar landbouwprodukten. Op de horizontale as staat een indeling van landbouwstelsels naar de mate van menging tussen gewassen en vee (Hfdst. 2.2). De combinatie van de relatieve toegang tot produktiefaktoren en de vraag naar dierlijke produkten worden hier de randvoorwaarden (het resource / demand pattern) genoemd. Deze indeling blijkt belangrijk te zijn om de logica van systeemontwikkeling te begrijpen en om, per subklasse uit de matrix, het nut van strovoederingsmethoden aan te geven (Hfdst. 2.3). Uit de discussie blijkt, dat deze indeling ook een basis vormt voor een verklaring van systeemgedrag (Hfdst. 6).

Tijdens de pogingen om systeemgedrag te verklaren, werd het steeds duidelijker dat er parallellen bestaan tussen ontwikkeling en vorming (morphogenese) van systemen. Beide termen, vorming en randvoorwaarden, worden gebruikt in de meer fundamentele discussie over systeemontwikkeling in Hfdst. 6. De indeling gebaseerd op de randvoorwaarden blijkt het meer traditionele onderscheid tussen tropische en gematigde landbouwstelsels te vervangen door een meer toepasselijke voor dit proefschrift, nl. een indeling, gebaseerd op het onderscheid tussen open en gesloten systemen (Hfdst. 2.2 en 5.1).

Het literatuuroverzicht van Hfdst. 2.2 laat verder zien dat ontwikkeling niet altijd synoniem is met verbetering. Ontwikkeling blijkt vaak een door de nood gedreven antwoord te zijn op veranderende grensvoorwaarden. Praktisch gesproken betreft de "nood" hier de behoefte aan "goederen" zoals: voedsel, kleding, woning, transport, en ontspanning. Op een abstracter niveau betreft de "nood" een behoefte aan verschillende vormen van energie. Overigens, de uiteindelijke behoefte wordt niet alleen bepaald door het aantal mensen, maar door het produkt van het aantal mensen en de individuele consumptie.

De veranderende behoefte dwingt een landbouwstelsel tot het invoeren van en het zoeken naar zogenaamde gedwongen vernieuwingen (induced innovations). Het zoeken naar en de toepassing van nieuwe strovoederingsmethoden is daarvan een typisch voorbeeld. Evenals bij veel andere technologieën, kunnen deze vernieuwingen een - korte termijn - verbetering zijn voor het (sub)systeem waarin ze worden toegepast. Echter, zij blijken vaak te berusten op het gebruik van hulpbronnen uit een ander (sub)systeem. In dit verband wordt het Simon effect ingevoerd: het verkrijgen van een beter resultaat in het ene (sub)systeem op kosten van een ander (sub)systeem (Hfdst. 2.2).

Componentonderzoek van strovoedering

Ten einde het nut van enkele van de vele methoden van strovoedering te voorspellen, werd een aantal voederproeven uitgevoerd. Dit soort werk heet "component onderzoek", een vorm van werken, die op zich ondersteunend kan zijn voor FSR. De proeven hadden tot doel het gebruik van ontsloten stro te vergelijken met dat van niet ontsloten stro, al dan niet in combinatie met bijvoedering (Hfdst. 3.1-3.4). Om de resultaten van deze proeven algemeen toepasbaar te maken, werden de proeven in toenemende mate zo opgezet, dat er basisparameters uit de veevoeding werden bepaald, nl.: de verteerbaarheid en opname van voeders. De discussie laat zien dat deze parameters eigenlijk de energiestromen in systemen vertegenwoordigen (Hfdst. 6).

De proeven hebben laten zien, samen met literatuurgegevens, dat stro-ontsluiting met ureum leidt tot een hogere stro-opname en een betere beschikbaarheid van verteerbare energie van het stro voor het vee. Het ruw eiwit gehalte van het stro neemt ook toe, maar dat is vanuit voedingsoogpunt minder belangrijk en het kan eenvoudiger worden bereikt met bijvoedering. Bijvoedering zelf kan ook leiden tot betere verteerbaarheid en hogere stro-opname, het zgn. katalytisch effect. Het resultaat daarvan lijkt echter geringer en minder zeker dan bij de stro-ontsluiting.

Het nut van stro-ontsluiting

De economische haalbaarheid van stro-ontsluiting werd berekend met een aantal denkexperimenten, in dit geval vereenvoudigde economische berekeningen, gebaseerd op parameters uit de literatuur (Hfdst. 4.1). De uitkomsten werden getoetst aan sociaal-economische en voedertechnische overwegingen (Hfdst. 4.2 en 5.1) en aan de resultaten van de voederproeven (Hfdst. 3.1-3.4). De conclusie is, dat stro-ontsluiting een typische "niche" vernieuwing is. Met andere woorden, het antwoord op de eerste vraag van dit proefschrift is, dat het nut en de wijze van strovoedering in grote mate afhangt van de situatie in en rond het landbouwstelsel, oftewel van de randvoorwaarden. Meer specifiek gezegd, het antwoord op de eerste vraag van dit proefschrift is, dat stro-ontsluiting vooral zinvol is in stelsels, waar

- genoeg stro beschikbaar is, d.w.z. waar stro goedkoop is in vergelijking tot andere veevoeders
- geen hoge produkties per dier nodig zijn
- toegang is tot produktiemiddelen (inputs), in ruil voor produkten van de dieren (out-puts).

Achteraf gezien is het niet zo verwonderlijk dat er ecologische termen worden gebruikt, zoals niche oplossingen en systeem morphogenese. In het bijzonder in de laatste fase van het werk aan dit proefschrift werd het steeds duidelijker, dat de ontwikkeling van landbouwstelsels lijkt op het door Darwin beschreven proces van het ontstaan van de soorten. M.a.w., de vorming van bedrijfsstelsels lijkt op een evolutie van soorten, waarin het sterkste overleeft. Door de grote variatie aan omstandigheden (niches) zijn meerdere bedrijfsstelsels mogelijk, uiteraard zijn het hierbij de randvoorwaarden van het systeem die de vorming ervan bepalen.

Open en gesloten systemen, damning objectives en het Simon-effect

In dit verband was het nodig om nog meer aandacht te besteden aan het belang van randvoorwaarden voor het nut van methoden van (stro)voedering. Daartoe wordt het gedrag van relatief gesloten landbouwstelsels in LEIA (= lage input landbouw) vergeleken met dat van relatief open landbouwstelsels in HEIA (= hoge input landbouw) (Hfdst. 5.1). Een belangrijk verschil tussen beide systemen blijkt dan te zijn dat de meer gesloten systemen geneigd zijn hun doelen aan te passen aan de beschikbaarheid van produktiemiddelen (inputs), terwijl de open systemen ertoe neigen om de beschikbaarheid van inputs aan te passen aan hun doelen.

In de planning van landbouwontwikkeling wordt veelal gebruik gemaakt van de "open systeem" benadering. Men gaat er tot op bepaalde hoogte vanuit, dat produktiemiddelen aangeschaft kunnen worden, naar gelang ze nodig zijn om het gestelde doel te bereiken. Om te onderzoeken wat er gebeurt in een "gesloten systeem" benadering werden er opnieuw een aantal denkexperimenten gedaan. Het resultaat laat zien, dat bij zulke gesloten omstandigheden, produktiedoelen (voor subsystemen) aangepast moeten worden aan de hulpbronnen. Door niet à priori uit te gaan van hoge individuele produkties per dier (subsysteem), is het soms mogelijk een hogere produktie uit het totale systeem te halen (Hfdst. 5.2). Een produktiedoel van een subsysteem, dat te hoog en te rigide is gesteld, wordt hier tot een zogenaamd damning objective. Dit betekent dat zo'n doelstelling onmogelijk is, of dat ze alleen bereikt kan worden door ontbrekende hulpbronnen uit andere systemen te betrekken. Zo'n damning objective kan zelfs het gebruik van op het bedrijf aanwezige laagwaardige voeders verhinderen. Het resultaat is dat zo'n (laagwaardig) produktiemiddel (in dit geval voer) verwordt tot afval, met vervuiling als gevolg. Bovendien blijkt in gesloten systemen de totale hoeveelheid produkt, bijvoorbeeld in de vorm van melk, mest en vlees, vergroot te kunnen worden door hoog en laag producerende dieren te combineren. Dit bepleit enerzijds het toepassen van verschillende en flexibele criteria voor verschillende subsystemen, anderzijds zet het vraagtekens bij het succes van hoge produkties van individuele subsystemen en verregaande standaardisatie van b.v. koeien of bedrijven.

Deze principes van het "damning objective" en het "Simon effect" zijn vermoedelijk complementair en toepasbaar voor systemen op ieder niveau, een cel, een koe, een landbouwstelsel of een samenleving. De relatie tussen damning objectives en de problemen van sociale ongelijkheid door ontwikkeling, zoals bedoeld met het Simon effect, lijkt hiermee te zijn gelegd. In plaats van de nadruk te leggen op de ontwikkeling van individueel hoog produktieve subsystemen, zou een systeem zich kunnen richten op het gebruik van zogenaamde communale ideotypes (Hfdst. 6).

De drijvende kracht en de vorm van systemen

In de discussie worden de dusver verkregen conclusies getoetst en enkele losse einden van het betoog tot nu tot verklaard (Hfdst. 6). Het gaat hierbij met name om het nut van strovoederingsmethoden in verschillende bedrijfsstelsels, de noodzaak van verschillende maatstaven voor het meten van systeemsucces, het Simon effect en het damning objective. Verder wordt althans een voorlopig antwoord gegeven op de tweede vraag van dit proefschrift, *de rol van stro in de drijvende kracht en de vorming van bedrijfsstelsels.* Het gebruik van begrippen uit de thermodynamica en informatietheorie is daarbij van groot belang. Centraal staat de tweede wet van de thermodynamica, die de drijvende kracht van systemen verklaart door te zeggen dat alle systemen, wanneer ze aan zichzelf worden overgelaten, neigen tot een toestand van maximale entropie oftewel afwezigheid van orde. Zowel energie als informatie zijn daarom vereist om een systeem te organiseren en in stand te houden. Het doet er in dit verband waarschijnlijk weer niet toe of het systeem een levende cel, een dier, een landbouwstelsel of een samenleving is. De analogieën tussen systeemgedrag, zoals herkend in het overzicht van FSR, blijken hierdoor nog relevanter te zijn dan eerst werd gedacht in Hfdst. 2.1.

De rol van stro en methoden van strovoedering in de vorming van bedrijfsstelsels valt op deze wijze, althans ook voor een deel te verklaren. Als de voedingswaarde van veevoer wordt uitgedrukt in termen van voederopname en energieverteerbaarheid, dan is het aannemelijk te maken dat voeders van een groter verteerbaarheid en opname, in staat zijn om complexere systemen te vormen en in stand te houden. Bij voldoend hoge stromen (fluxen) van energie en informatie zijn de complexere systemen meestal in staat tot een hogere bruto opbrengst van b.v. melk en vlees. Overigens gaat een hogere opbrengst niet noodzakelijkerwijs gepaard met een hogere energie efficiëntie van het systeem. In dit verband is het niet verwonderlijk dat ontsloten stro, als voer met een matige energieflux, vooral zinvol is voor dieren met een matige produktie van melk en vlees. Lagere en hogere voerkwaliteit zullen systemen mogelijk maken die navenant eenvoudiger of primitiever, dan wel ingewikkelder of meer ontwikkeld zijn. In meer ontwikkelde bedrijfsstelsels met meer produktieve dieren neemt het nut van stro als veevoer af: een voer dat nuttig is in het ene systeem kan dus "afval" blijken te zijn in het andere.

In deze zin is het duidelijk dat vee, op een goede wijze gevoerd en gehouden, een rol speelt in de beheersing van systemen en als zodanig rijkdom vertegenwoordigt. Het gebruik van inzichten uit verschillende vakgebieden is een typische randvoorwaarde voor het doen van FSR. Het zal ook kunnen bijdragen tot verder begrip en kwantificering van de rol van vee en stro in systeembeheersing.

CURRICULUM VITAE

Johannes Bouwe Schiere werd op 12 april 1949 in Leeuwarden geboren en groeide op in Surhuisterveen. Hij behaalde het HBS-B diploma in 1967 aan het Drachtster Lyceum te Drachten. Alvorens met de studie aan de toenmalige Landbouw Hogeschool te Wageningen te beginnen, werkte hij een jaar op boerenbedrijven in de Verenigde Staten. Tijdens zijn studie tropische veeteelt werd de praktijktijd doorgebracht in Iran en Indonesië. Op het eind van zijn studie maakte hij deel uit van een interdisciplinaire werkgroep op het Internationaal Agrarisch Centrum te Wageningen: *"de kleine boer en de ontwikkelings- samenwerking"*. Na het behalen van het ingenieurs diploma in 1975, werkte hij achtereen- volgens voor DGIS in een project voor melkvee- en graslandontwikkeling in Tarapoto, Peru (1976-1979), voor NUFFIC bij de opbouw van de veeteeltfaculteit van de Universitas Brawijaya in Malang, Indonesië (1980-1983) en voor DGIS in een onderzoeks / voorlich- tingsprojekt op het gebied van strovoedering in Sri Lanka (1983-1986). Vanaf juli 1986 tot december 1994 was hij in deeltijd betrokken bij de sektie Tropische Veeteelt van de Landbouw Universiteit als coördinator van door DGIS gefinancierde onderzoeks / voorlichtingsprojekten in Indonesië en India. Vooral in India hield hij zich bezig met onderzoek en voorlichting over veevoeding in het kader van landbouwstelsel onderzoek (Farming Systems Research). Een deel van de vrije tijd werd besteed aan het schrijven van dit proefschrift. In de rest van zijn tijd nam hij deel aan adviseringsmissies voor FAO, DGIS en particuliere organisaties in Latijns-Amerika, Azië en Afrika. In dat verband is hij betrokken geweest bij het opstellen van beleidsdocumenten en publikaties, vooral op het gebied van duurzaamheid in gemengde vee en akkerbouwsystemen. Bij dit alles is hij vooral geboeid door de mogelijkheden van interdisciplinaire samenwerking, op technisch zowel als sociaal-economisch gebied. Sinds begin 1995 werkt hij bij de sektie Dierlijke Produktie Systemen van de Landbouw Universiteit.